Analytical Chemistry of Complex Matrices

Analytical Chemistry of Complex Matrices

W. Franklin Smyth
University of Ulster

WILEY TEUBNER

A Partnership between John Wiley & Sons and B. G. Teubner Publishers

Chichester · New York · Brisbane · Toronto · Singapore · Stuttgart · Leipzig

Copyright 1996 by John Wiley & Sons Ltd and B. G. Teubner

John Wiley & Sons Ltd
Baffins Lane
Chichester
West Sussex PO19 1UD
England

B.G. Teubner
Industriestraße 15
70565 Stuttgart (Vaihingen)
Postfach 80 10 69
70510 Stuttgart
Germany

National   Chichester (01243) 779777
International   +44 1243 779777

National   Stuttgart (0711) 789 010
International   +49 711 789 010

All rights reserved

No part of this book may be reproduced by any means, or transmitted, or translated into a machine language without the written permission of the publisher.

Designations used by companies to distinguish their products are often claimed as trademarks. In all instances where John Wiley & Sons Ltd and B.G. Teubner are aware of a claim, the product names appear in initial capital or all capital letters. Readers, however, should contact the appropriate companies for more complete information regarding trademarks and registration.

Other Wiley Editorial Offices

John Wiley & Sons, Inc., 605 Third Avenue,
New York, NY 10158-0012, USA

Brisbane · Toronto · Singapore

Other Teubner Editorial Offices

B.G. Teubner, Verlagsgesellschaft mbH, Johannisgasse 16,
D-04103 Leipzig, Germany

Library of Congress Cataloging-in-Publication Data
Smyth, W. Franklin.
  Analytical chemistry of complex matrices / W. Franklin Smyth.
      p.   cm.
  Includes bibliographical references and index.
  ISBN 0-471-96316-X (cloth / alk. paper)
  1. Chemistry, Analytic. I. Title.
  QD75.2.S69   1996   95-44666
  543—dc20   CIP

Die Deutsche Bibliothek - CIP-Einheitsaufnahme
Smyth, W. Franklin:
Analytical chemistry of complex matrices / W. Franklin Smyth.
–Stuttgart ; Leipzig ; Teubner : Chichester : New York : Brisbane : Toronto : Singapore : Wiley, 1996
  ISBN 3-519-02116-1 (Teubner)
  ISBN (falsch) 0-471-96316-X (Wiley)

British Library Cataloguing in Publication Data

A catalogue record for this book is available from the British Library

ISBN Wiley 0 471 96316 X

ISBN Teubner 3-519-02116-1

Typeset in 10/12pt Times by Techset Composition Ltd, Salisbury, Wiltshire
Printed and bound in Great Britain by Biddles Ltd, Guildford, Surrey

This book is printed on acid-free paper responsibly manufactured from sustainable forestation, for which at least two trees are planted for each one used for paper production.

Dedication

Analytical Chemistry of Complex Matrices is dedicated to my mentors in analytical chemistry over the years who, through my reading of their contributions to the literature and, in some cases, meeting them in person, have enhanced my understanding of analytical chemistry and its application to the complex matrices of today's world.

Contents

INTRODUCTION	1
CHAPTER 1: HISTORICAL BACKGROUND	12
CHAPTER 2: UNIT PROCESSES OF ANALYTICAL PROCEDURES	18
2.1 Unit Process No. 1. Definition of the Problem	19
2.2 Unit Process No. 2. Choice of Method	19
2.3 Unit Process No. 3. Obtaining a Representative Sample and its Measurement	37
2.3.1 Solid Materials	37
2.3.2 Liquid Materials	39
2.3.3 Gaseous Materials	40
2.4 Unit Process No. 4. Preliminary Treatment of Sample	42
2.5 Unit Process No. 5. Separation of Analyte(s) from Interferences and Each Other	45
2.5.1 Separation by Masking	45
2.5.2 Separation by Precipitation	45
2.5.3 Separation by Solvent Extraction	46
2.5.4 Separation by Solid-Phase Extraction	48
2.5.5 Separation by Chromatography	49
2.5.6 Separation by Electrophoresis	49
2.6 Unit Process No. 6. Measurement	50
2.7 Unit Process No. 7. Statistical Assessment of Measurements	51
2.8 Unit Process No. 8. Calculation of Analytical Result and Solution to Problem	55
2.8.1 Quantitative Analysis	55
2.8.1.1 Calibration Curves	55
2.8.1.2 Standard Addition Method	58
2.8.1.3 Internal Standard Method	59

		2.8.1.4 Internal Normalization Method	60

- 2.8.2 Qualitative Analysis. 61
- 2.9 Rule of Computers and Microprocessors in Modern Analytical Methods. 61
  - 2.9.1 Instrument Operation 61
  - 2.9.2 Data Recording and Storage 64
  - 2.9.3 Data Processing and Analysis 64
  - 2.9.4 Validation Testing. 65
  - 2.9.5 Software for Method Development 67
- 2.10 Automation of Unit Processes 70
  - 2.10.1 Automation of Repetitive Analysis 70
  - 2.10.2 Continuous On-Line Monitoring 75
  - 2.10.3 Laboratory Robotics 76
  - 2.10.4 Application of Robotics to Dissolution Tests 77
  - 2.10.5 Application of Robotics to Drug Determination in Biological Fluids. 77

CHAPTER 3: SELECTED ANALYTICAL PROBLEMS INVOLVING INORGANIC ANALYTES WHICH CONTAIN ELEMENTS FROM GROUPS IA–VIIIA AND THE LANTHANIDES — 79

- 3.1 Determination of Sodium and Potassium in Mineral Water by Flame Emission Spectrometry 79
  - 3.1.1 Summary. 79
  - 3.1.2 Introduction 79
  - 3.1.3 Procedure 80
- 3.2 Determination of Water Hardness, i.e. Total Calcium and Magnesium, by EDTA Titration 80
  - 3.2.1 Summary. 80
  - 3.2.2 Introduction 81
  - 3.2.3 Procedure 82
  - 3.2.4 Calculation. 82
- 3.3 Gravimetric Determination of Cerium in Ore. 83
  - 3.3.1 Summary. 83
  - 3.3.2 Introduction 83
  - 3.3.3 Procedure 83
- 3.4 Spectrophotometric Determination of Titanium in Rock 84
  - 3.4.1 Summary. 84
  - 3.4.2 Introduction 84
  - 3.4.3 Procedure 85
  - 3.4.4 Calculation. 86
- 3.5 Determination of Vanadium(V) in Sea Water by Adsorptive Stripping Voltammetry 87
  - 3.5.1 Summary. 87

Contents ix

	3.5.2 Introduction	87
	3.5.3 Procedure	88
3.6	Determination of Trace Concentrations of Molybdenum in Water Samples by Solvent Extraction of a Neutrally Charged Chelate Followed by Atomic Absorption Spectrometry	89
	3.6.1 Summary	89
	3.6.2 Introduction	90
	3.6.3 Procedure	90
3.7	Differential Thermal Analysis (DTA) of a Complex Manganese Compound $Mn(PH_2O_2)H_2O$	90
	3.7.1 Summary	90
	3.7.2 Introduction	90
	3.7.3 Procedure	91
3.8	Determination of Iron in an Ore by Redox Titration with Potassium Permanganate	92
	3.8.1 Summary	92
	3.8.2 Introduction	92
	3.8.3 Procedure	93
	3.8.4 Calculation	94
3.9	Determination of Cobalt in Soil Samples by Visible Spectrophotometry	94
	3.9.1 Summary	94
	3.9.2 Introduction	94
	3.9.3 Procedure and Calculation	94
3.10	Gravimetric Determination of Nickel in Steel	96
	3.10.1 Summary	96
	3.10.2 Introduction	96
	3.10.3 Procedure	97

CHAPTER 4: SELECTED ANALYTICAL PROBLEMS INVOLVING INORGANIC AND ORGANOMETALLIC ANALYTES WHICH CONTAIN ELEMENTS FROM GROUPS IB–VIIB   98

4.1	Determination of Trace Concentrations of Copper in the Aqueous Environment by Stripping Voltammetry	98
	4.1.1 Summary	98
	4.1.2 Introduction	98
	4.1.3 Procedure	99
	4.1.4 Calculation	101
4.2	Determination of Organomercury Compounds in Fish Samples by HPLC–Cold Vapour Atomic Absorption Spectrometry	102
	4.2.1 Summary	102
	4.2.2 Introduction	102
	4.2.3 Procedure	104

- 4.3 Determination of Zinc in a Pharmaceutical Formulation by Ion-Exchange Separation and Complexometric Titration 104
  - 4.3.1 Summary.................................. 104
  - 4.3.2 Introduction 105
  - 4.3.3 Procedure 108
- 4.4 Visible Spectrophotometric Determination of Boron in Plants 108
  - 4.4.1 Summary.................................. 108
  - 4.4.2 Introduction 109
  - 4.4.3 Procedure 110
  - 4.4.4 Calculation................................ 110
- 4.5 Determination of Organolead Compounds in Air Samples by Gas Chromatography–Atomic Absorption Spectrometry 112
  - 4.5.1 Summary.................................. 112
  - 4.5.2 Introduction 112
  - 4.5.3 Procedure 113
- 4.6 Determination of Nitrogen Dioxide in Air Samples by Sorbent Tube Collection–Colorimetry........................ 113
  - 4.6.1 Summary.................................. 113
  - 4.6.2 Introduction 115
  - 4.6.3 Procedure 123
  - 4.6.4 Calculation................................ 123
- 4.7 Determination of Arsenic in Hair Samples by Neutron Activation Analysis and in Wallpaper by X-ray Fluorescence Spectrometry..... 125
  - 4.7.1 Summary.................................. 125
  - 4.7.2 Introduction 125
  - 4.7.3 Procedure 127
- 4.8 Determination of Antimony in Liver Samples using Hydride Generation–Atomic Absorption Spectrometry 130
  - 4.8.1 Summary.................................. 130
  - 4.8.2 Introduction 130
  - 4.8.3 Procedure 131
- 4.9 Determination of Selenium in Natural Waters by Spectrofluorimetry .. 132
  - 4.9.1 Summary.................................. 132
  - 4.9.2 Introduction 132
  - 4.9.3 Procedure and Calculation 133
- 4.10 Determination of Fluoride in Potable Water by Ion-Selective Electrode Measurements 135
  - 4.10.1 Summary.................................. 135
  - 4.10.2 Introduction 135
  - 4.10.3 Procedure 137
  - 4.10.4 Calculation................................ 137
- 4.11 Multielement Analysis of a Biological Fluid by Inductively Coupled Plasma Mass Spectrometry (ICP-MS).................... 139
  - 4.11.1 Summary.................................. 139

	4.11.2	Introduction	140
	4.11.3	Procedure	141

CHAPTER 5: SELECTED ANALYTICAL PROBLEMS INVOLVING ORGANIC ANALYTES WHICH ARE THE MAJOR OR MINOR CONSTITUENTS OF A SAMPLE — 143

5.1 Comparison of Analytical Methods Based on Visible Spectrophotometry, Solvent Extraction–UV Spectrophotometry, Voltammetry, NMR and HPLC for the Determination of the Active Constituents of Analgesic Formulations 143
   5.1.1 Summary 143
   5.1.2 Visible Spectrophotometric Method 144
      5.1.2.1 Introduction 144
      5.1.2.2 Procedure 144
      5.1.2.3 Calculation 145
   5.1.3 Solvent Extraction–UV Spectrophotometric Method 146
      5.1.3.1 Introduction and Procedure 146
   5.1.4 Voltammetric Method 147
      5.1.4.1 Introduction 147
      5.1.4.2 Procedure 147
   5.1.5 NMR Method 147
      5.1.5.1 Introduction 147
      5.1.5.2 Calculation 149
   5.1.6 HPLC Method 149
      5.1.6.1 Introduction 149
      5.1.6.2 Procedure 150
5.2 Stability-Indicating High-Performance Liquid Chromatographic Assay for Oxazepam Tablets and Capsules 150
   5.2.1 Summary 150
   5.2.2 Introduction 150
   5.2.3 Procedure 152
5.3 Separation of Water-Soluble Vitamins by Capillary Zone Electrophoresis and Micellar Electrophoretic Capillary Chromatography (MECC) 154
   5.3.1 Summary 154
   5.3.2 Introduction 154
   5.3.3 Procedure 155
5.4 Air-Segmented Continuous-Flow Visible Spectrophotometric Determination of Cephalosporins in Drug Formulations by Alkaline Degradation to Hydrogen Sulphide and Formation of Methylene Blue — 157
   5.4.1 Summary 157
   5.4.2 Introduction 157
   5.4.3 Procedure 157

5.5	Use of IR Spectrometry to Identify an Active Raw Material, Procaine Penicillin G, Used in the Pharmaceutical Industry	157
	5.5.1 Summary	157
	5.5.2 Introduction	158
	5.5.3 Procedure	160
5.6	Qualitative and Quantitative Analysis of a Polymeric Material using Thermogravimetric Analysis (TGA)	161
	5.6.1 Summary	161
	5.6.2 Introduction	164
	5.6.3 Procedure	164

CHAPTER 6: ORGANIC TRACE ANALYSIS OF LOW MOLECULAR WEIGHT ANALYTES IN ENVIRONMENTAL SAMPLES AND BIOLOGICAL MATERIALS   165

6.1	Determination of Aromatic and Aliphatic Isocyanates in Workplace Atmospheres by HPLC	165
	6.1.1 Summary	165
	6.1.2 Introduction	165
	6.1.3 Procedure	165
	6.1.4 Calculation	167
6.2	Determination of Triazine Pesticide Residues in Environmental Samples by Enzyme-Linked Immunosorbent Assay (ELISA)	170
	6.2.1 Summary	170
	6.2.2 Introduction	170
	6.2.3 Procedure	171
6.3	Analysis of Alcoholic Beverages	173
	6.3.1 Summary	173
	6.3.2 Introduction	173
	6.3.3 Procedure	174
6.4	Determination of Contaminating Antibiotic Trace Concentrations in Dairy Feedstuffs by GC–MS	176
	6.4.1 Summary	176
	6.4.2 Introduction	176
	6.4.3 Procedure	176
6.5	Extraction of Cocaine and its Metabolites from a Urine Sample by Solid-Phase Extraction	179
	6.5.1 Summary	179
	6.5.2 Introduction	179
	6.5.3 Procedure	183
6.6	Determination of Nitroglycerin, 2,4-Dinitrotoluene and Diphenylamine in Gunshot Residue by High-Performance Liquid Chromatography with Electrochemical Detection (HPLC–ED)	184
	6.6.1 Summary	184

	6.6.2	Introduction	184
	6.6.3	Procedure	185
6.7	Determination of Prozac (Fluoxetine) and its Demethylated Metabolite in Serum by HPLC with Fluorescence Detection		187
	6.7.1	Summary	187
	6.7.2	Introduction	187
	6.7.3	Procedure	188
6.8	Direct Determination of Thioamide Drugs in Biological Fluids by Cathodic Stripping Voltammetry		190
	6.8.1	Summary	190
	6.8.2	Introduction	190
	6.8.3	Procedure	191
	6.8.4	Calculation	193

CHAPTER 7: ANALYSIS OF HIGH MOLECULAR WEIGHT ANALYTES — 194

7.1	Analysis of Abnormal or Variant Haemoglobins by HPLC–Electrospray MS		194
	7.1.1	Summary	194
	7.1.2	Introduction	194
	7.1.3	Procedure	196
7.2	Determination of Insulin by Radioimmunoassay (RIA)		199
	7.2.1	Summary	199
	7.2.2	Introduction	199
	7.2.3	Procedure	200
	7.2.4	Calculation	200
7.3	Fluorescence of Europium Chelates and its Application to Fluoroimmunoassay of Prostate-Specific Antigen (PSA) in Human Serum		201
	7.3.1	Summary	201
	7.3.2	Introduction	201
	7.3.3	Procedure	203

REFERENCES — 205

INDEX — 209

W. Franklin Smyth, CChem, BSc, PhD, DSc, FRSC, FICI, born in 1945 in Belfast, Northern Ireland, completed his PhD at The Queen's University of Belfast in 1970 under the supervision of Professor G. Svehla and Professor P. Zinman in the Department of Professor Cecil L. Wilson.

From 1970 to 1979 he lectured in analytical chemistry at Chelsea College, University of London, and from 1980 to 1982 at University College, Cork. Following a period in industry with Norbrook Laboratories, Newry, and a visiting professorship to the Department of Pharmacy, The Queen's University of Belfast, Dr Smyth was contracted by the University of Zambia, Lusaka and supported by the Overseas Development Administration UK as Professor of Chemistry for the period 1987 to 1989.

Dr Smyth joined the University of Ulster in 1990 and currently teaches and carries out research in the field of analytical chemistry of molecules of biological significance in complex matrices, which is essentially the subject of this book. Some 100 papers and reviews, two edited volumes on electroanalysis and a specialist monograph *Voltammetric Determination of Molecules of Biological Significance*, published by John Wiley & Sons, 1992, have been published by Dr Smyth. He has also held visiting appointments in universities in Aarhus, Buffalo, Turku and Haifa.

Dr Smyth is married and has five daughters.

ACKNOWLEDGEMENTS

The author wishes to thank Carol Millar, Secretary of the School of Applied Biological and Chemical Sciences, University of Ulster, Coleraine, Margaret Avenell also of the School of Applied Biological and Chemical Sciences, and Anne Hayes and Caroline Adams, Faculty Research Support Group, Science Faculty, University of Ulster, Coleraine, for their secretarial support in the preparation of the manuscript. Sharon Dornan of the Audio Visual Services Department, University of Ulster, Coleraine, is thanked for preparing the figures used in this volume.

INTRODUCTION

Analytical Chemistry of Complex Matrices is an introduction for the analytical scientist to analytical problems concerned with the identification and determination of organic, inorganic and organometallic analytes in complex matrices of topical importance such as the atmosphere, factory air, natural waters, industrial effluents, drinking water, plants, soils, minerals, foods and industrial products. It is assumed that the analytical scientist is already familiar with the theory and practice of a wide range of analytical methods and techniques such as are available in the many textbooks of analytical chemistry/science, as outlined by Locke and Grossman.[1] It is intended that this text should fill a void in analytical chemistry/science, since there is currently a distinct lack of educational literature dedicated to the evaluation and solution of the aforementioned analytical problems. This text has therefore been written with this end in mind and is based on the author's experiences in teaching and research in analytical chemistry/science to final-year undergraduate and postgraduate students in universities in London, Aarhus, Buffalo, Cork, Turku, Haifa, Belfast, Lusaka and Coleraine.

The text covers a wide range of analytical techniques from classical gravimetry and titrimetry, which illustrate basic chemical principles to the undergraduate, to instrumental techniques such as UV-visible spectrophotometry, IR spectrometry, spectrofluorimetry, gas chromatography, potentiometry and polarography used by undergraduates and postgraduates in most university teaching and research laboratories. State-of-the-art instrumental techniques such as inductively coupled plasma mass spectrometry and high-performance liquid chromatography with electrospray mass spectrometric detection, increasingly used in large industrial organisations and well funded government institutions and universities, are also covered. The text is therefore not only of value to final-year undergraduate and postgraduate science students whose courses and research projects contain qualitative and quantitative analyses of complex matrices, but also to analytical scientists in government and industrial laboratories who are inexperienced with particular measurements.

The text commences with a relatively brief historical background to analytical chemistry/science (Chapter 1). Chapter 2 considers the unit processes of an analytical method, i.e. definition of the problem, choice of method, obtaining a representative sample and its measurement, preliminary treatment of sample, separation of analyte(s) from interferences and each other, measurement, statistical assessment of measurements, calculation of analytical result and solution to the problem. Worked calculations are included under the unit process headings where appropriate. These are seen as being of particular educational value in training the novice analytical scientist to deal effectively with calculations. The important role of computers and microprocessors in modern analytical methods and the automation of unit processes within an analytical method are also discussed in this chapter.

Chapters 3 and 4 are concerned with selected analytical problems involving inorganic and organometallic analytes. Analytical problems are chosen so as to give examples of analyses involving a wide spread of elements in the Periodic Table (Figure 1), starting with Groups IA and IIA, through the transition elements to the B Groups and concluding with Group VIIB, the halogens. Table 1 surveys the properties of these selected elements (e.g., isotopes, minerals, naturally occurring compounds, toxicity) and particularly includes information on the typical concentrations (mg kg^{-1} for liquid and solid samples and ng m^{-3} for air samples) in which they are found in a variety of complex matrices. This latter information then forms the basis of some of the analytical problems encountered in Chapters 3 and 4 e.g., vanadium in sea water (Section 3.5), cobalt in soil (Section 3.9) and boron in plants (Section 4.4). Many of the problems are concerned with single element analysis (e.g. cerium, molybdenum, iron, cobalt, nickel, zinc, fluorine), but attention is also paid to multielement analysis [e.g. lead, barium, tin, iodine, sodium, calcium, magnesium, iron, phosphorus, copper, zinc, rubidium, bromine, aluminium and manganese (Section 4.11)] and to both inorganic speciation [e.g., copper (Section 4.1) and selenium (Section 4.9)] and organometallic speciation [e.g., organomercury compounds (Section 4.2) and organolead compounds (Section 4.5)]. The problems in Chapters 3 and 4 are also chosen as is the case in Chapters 5–7 on organic analytes, to give a variety of both complex matrices investigated and the analytical methods and techniques used in problem solution. Chapter 5 covers the analysis of major (100–1%) and minor (1–0.01%) organic constituents of a sample, Chapter 6 organic trace analysis (10^2–10^{-4} ppm) of low molecular weight analytes in environmental samples and biological materials and Chapter 7 the analysis of high molecular weight analytes. Many of these analytical problems in Chapters 5–7 are concerned with the analysis of multicomponent mixtures (e.g. isocyanates, ethyl acetate and higher alcohols, oxazepam and degradation products, haemoglobins and water-soluble vitamins). Table 2 gives a complete list of the analytical problems encountered in this text that are concerned with inorganic, organometallic and organic analytes. It is tabulated in terms of the analytes to be identified and/or determined, the complex matrices initially presented to the analytical scientist and the techniques used in the measurement steps. A summary which, where possible, compares and contrasts the chosen analytical method with alternative strategies is provided for each problem prior to an introduction to the problem and procedural details for the practical solution of the problem. Worked calculations are included for some of the problems since these are seen as being of particular educational value in the numerical aspects of problem solving. In addition, particular attention is paid to the underlying chemistry of each analytical method.

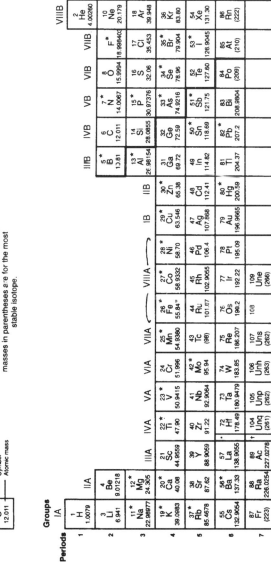
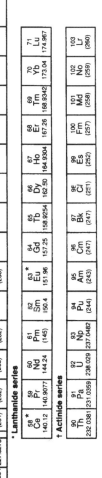

Figure 1 Selection of elements from the Periodic Table, indicated by asterisks, that are the analytes in the analytical problems in Chapters 3 and 4. Note that the fluorescence of europium chelates is dealt with in Chapter 7 since it is applied to the solution of an organic problem, namely the determination of prostate-specific antigen in human serum by fluoroimmunoassay

Table 1 Properties of selected elements and their concentrations in a variety of complex matrices[2] (some of these data have been updated by reference to the World Wide Web using the server at Sheffield University, i.e. http://www.shef.ac.uk/~chem/web-elements/; 1995 Mark Winter, Department of Chemistry, University of Sheffield, Sheffield S3 7HF)

Element	Stable isotopes (fractional abundance); natural radioisotopes ($t_{1/2}$); minerals; naturally occurring compounds	Toxicity	Typical concentrations (mg kg^{-1})									Notes
			Limestone	Whole human blood[a]	Soil	Sea Water	Fresh Water	Air[b]	Edible vegetables[c]	Mammal muscle[c]	Marine fish[c]	
SODIUM Na (AW 22.99)	^{23}Na (1); ^{22}Na (2.58 yr); NaCl, albite, nepheline	Not very toxic; 320 mg day^{-1} is lethal to rats	1300	1970	5000	11050	6.3	7–7000	2–2400	2600–7800	8000	Functions: electrochemical and catalytic
POTASSIUM K (AW 39.0983)	^{39}K (0.931); ^{40}K (1.3 Gyr); feldspars, micas, KCl	Scarcely toxic except by intravenous injection to mammals	3100	1620	14000	416	2.3	0.3–40000	1000–68000	16000	15000	Essential to all organisms
CALCIUM Ca (AW 40.08)	^{40}Ca (0.970), also ^{42}Ca, ^{43}Ca, ^{44}Ca, ^{46}Ca, ^{48}Ca; CaCO$_3$, CaSO$_4$, feldspar, apatite, Ca alginate, Ca pectate, enzymes	Relatively harmless to all organisms	340 × 10^3	60.5	15 × 10^3	422	15	0.5–7000	400–50 × 10^3	140–700	76–20000	Functions: hard tissues, cell walls, electrochemical messenger in nerve and muscle response, etc.
MAGNESIUM Mg (AW 24.305)	^{24}Mg (0.787), ^{25}Mg (0.101), ^{26}Mg (0.112); olivine, talc, dolomite; Mg(OH)$_2$ (rare), chlorophylls, MgATP	Scarcely toxic except by intravenous injection	5800	37.8	5000	1326	4.1	1–11000	700–5600	900	1200	Essential to all organisms

Element	Isotopes/Minerals									Notes
CERIUM Ce (AW 140.12)	^{140}Ce(0.885), ^{142}Ce(0.111); ^{141}Ce(33d), ^{144}Ce(284d); monazite (CePO$_4$)	20	<0.002	50	1.2×10^{-6}	0.6×10^{-4}	0.002–20	0.01–0.05	<0.01	0.006–0.03	Ce is mildly toxic by ingestion but insoluble salts such as the oxalate are non toxic and doses of up to 500 mg were once prescribed to prevent travel sickness and morning sickness	No biological role but acts to stimulate metabolism
TITANIUM Ti (AW 47.90)	^{48}Ti (0.739) + several others; TiO$_2$, FeTiO$_3$, CaTiO$_3$, Sphene	300	0.054	5000	0.001	0.003	2–1000	<0.02–3	1–2	0.2	Relatively harmless	Av. daily intake for man is 0.8 mg
VANADIUM V (AW 50.9415)	^{51}V (0.998) + ^{50}V; ^{50}V (6 × 10^{15} yr); VS$_4$, calcium vanadate; amavanadine and haemovanadin	45	$<2 \times 10^{-4}$	90	1.5×10^{-3}	9×10^{-4}	0.0015–2000	0.001–0.5	0.002–0.02	0.03	10–40 mg.dm^{-3} toxic to plants, 0.25 mg day^{-1} toxic, 2–4 mg day^{-1} lethal to rats	Essential to all vertebrates; av. daily intake for man is 0.04–4 mg
MOLYBDENUM Mo (AW) 95.94	^{98}Mo (0.238), ^{96}Mo (0.165), ^{95}Mo (0.157) + others; ^{99}Mo (67 h); MoS$_2$; contained in a few Fe enzymes	0.16	0.001	1.2	0.01	6×10^{-4}	<0.2–10	0.03–5	0.02–0.07	1	0.5–2 mg dm^{-3} toxic to plants; 5 mg day^{-1} toxic, 50 mg day^{-1} lethal to rats	Essential to all organisms; used in N$_2$ fixation; High [Mo] in soil causes Cu deficiency in cattle

(continued overleaf)

Table 1 (*continued*)

Element	Stable isotopes (fractional abundance) ($t_\frac{1}{2}$); minerals; naturally occurring compounds	Toxicity	Typical concentrations (mg kg^{-1})									Notes
	radioisotopes ($t_\frac{1}{2}$); minerals; naturally occurring compounds		Limestone	Whole human blood[a]	Soil	Sea Water	Fresh Water	Air[b]	Edible vegetables[c]	Mammal muscle[c]	Marine fish[c]	
MANGANESE Mn (AW 54.9380)	^{55}Mn (1); ^{54}Mn (280 days); MnO$_2$, MnCO$_3$, MnSiO$_3$; 12 Mn proteins or enzymes	1–100 mg dm^{-3} toxic to plants; 10–20 mg day^{-1} toxic to rats	620	0.0016 –0.075	1000	$2 \times 10^{-3,d}$	7×10^{-3}	0.01– 900	0.3– 10^3	0.2–2.3	0.3–4.6	Essential to all organisms; av. daily intake for man is 0.4–10 mg
IRON Fe (AW 55.847)	^{56}Fe (0.917) + others; ^{55}Fe (2.6 yr); ^{59}Fe (45 days); FeS$_2$, Fe$_3$O$_4$, FeCO$_3$, Fe$_2$O$_3$, FeO$_2$H, Olivine; porphyrins, >70 metalloproteins and enzymes, ferritin	10–200 mg dm^{-3} toxic to plants; 200 mg day^{-1} toxic to man	17000	447	40000	3×10^{-3}	0.670	0.8– 14000	2–250	180	9–88	Essential to all organisms; accumulated by fish liver, red blood cells, certain lichens
COBALT Co (AW 58.9332)	^{59}Co (1); ^{60}Co (5.3 yr); CoS$_2$, CoAs$_2$ (with Cu); Vitamin B$_{12}$, some enzymes.	0.1–3 mg dm^{-3} toxic to plants; 500 mg day^{-1} toxic to man	0.1	2×10^{-4} –0.04	8	8×10^{-5}	1×10^{-4}	0.0008 –37	0.01– 4.6	0.005–1	0.006 –0.05	Essential element; low soil Co affects health of grazing mammals

Element	Isotopes; minerals; enzymes								
NICKEL Ni (AW 58.71)	^{58}Ni(0.679), ^{60}Ni(0.262) + others; ^{63}Ni(92 yr); NiS, garnierite; urease, hydrogenase	0.5–2 mg dm^{-3} toxic to plants; 50 mg day^{-1} toxic to rats	7	0.01–0.05	50	2×10^{-3}	3×10^{-4}	<1–120	0.02–4	1–2	0.1–4	carcinogen; has catalytic functions and stabilises coiled ribosomes
COPPER Cu (AW 63.55)	^{63}Cu(6.691), ^{65}Cu(0309); Cu$_2$S, CuS, CuFeS$_2$, CuO, Cu$_2$CO$_3$(OH)$_2$; 1 porphyrin and > 30 enzymes	0.5–8 mg dm^{-3} toxic to plants; 250 mg day^{-1} toxic to man	5.5	1.01	30	3×10^{-3}	7×10^{-3}	0.036–4900	4–20	10	0.7–15	Functions—pigment, antibiotic, O$_2$ transport, catalytic in redox reactions
ZINC Zn (AW 65.38)	^{64}Zn(0.489) + others; ^{65}Zn(245d); ZnS; > 80 metalloproteins, enzymes	60–400 mg dm^{-3} toxic to plants; 150–600 mg day^{-1} toxic to man	20	7.0	90	0.05	0.020	0.03–16000	1–160	240	9–80	Essential to all organisms
MERCURY Hg (AW 200.59)	196, 198, 199, 200, ^{201}Hg, 202, ^{204}Hg; ^{203}Hg (47 days); cinnabar (HgS); CH$_3$HgCl, Hg(CH$_3$)$_2$, C$_2$H$_5$HgCl	0.4 mg day^{-1} toxic, 150–300 mg day^{-1} lethal to man; metal vapour toxic to man at 44 mg m^{-3} (8 h); lethal to rabbit, LC$_{50}$ 29 mg m^{-3} (30 h)	0.16	0.0078	0.06	5×10^{-5}	0.7×10^{-4}	0.009–38	0.013–0.17	0.02–0.7	0.4	Accumulated by mammal kidney, reacts with SH groups and inhibits enzymes

(continued overleaf)

Table 1 (continued)

Element	Stable isotopes (fractional abundance), natural radioisotopes ($t_{1/2}$); minerals; naturally occurring compounds	Toxicity	Typical concentrations (mg kg^{-1})								Notes
			Limestone	Whole human blood^a	Soil	Sea Water	Fresh Water	Air^b	Edible vegetables^c	Mammal muscle^c	Marine fish^c	
BORON B (AW 10.81)	^{10}B (0.196) ^{11}B (0.804); borax, tourmaline; boromycin	1–5 mg dm^{-3} toxic to plants; 5 g/day toxic to man (as boric acid)	20	0.13	20	4.5	0.01	4	8–200	0.33–1	20	Accumulated by plants; one of its functions is that of an antibiotic
LEAD Pb (AW 207.2)	^{208}Pb (0.523) also $^{204, 206, 207}$Pb; ^{210}Pb (22 yr) and short-lived ^{211}Pb ^{212}Pb ^{214}Pb; PbS, PbCO$_3$, PbSO$_4$; Pb(CH$_3$)$_4$	3–20 mg dm^{-3} toxic to plants; 50 mg/day toxic, 10 g/day lethal to man	5.7	0.21	35 (contam.)	3×10^{-5}	3×10^{-3} (contam.)	0.6–13 200	0.2–20	0.23–3.3	0.001–15	Accumulated by mammalian bone and some lichens and plants
ARSENIC As (AW 74.9215)	^{75}As(1); ^{76}As (1.1 days); As$_2$S$_3$, As$_4$S$_4$, As, FeAs$_2$; Me$_3$As and HAsMe$_2$, MeAsO$_2$H, Me$_2$AsOH, Me$_3$AsO in urine	0.02–7.5 mg dm^{-3} toxic to plants; 5–50 mg day^{-1} toxic and 100–300 mg day^{-1} lethal to man	1	1.7×10^{-3} –0.09	6	2.3×10^{-3}	2×10^{-3}	1.5–53	0.01–1.5	0.007–0.09	0.2–10	Accumulated by mammalian hair and nails and some marine invertebrates

Element	Isotopes / compounds	Toxicity													Remarks
ANTIMONY Sb (AW 121.75)	¹²¹Sb(0.572), ¹²³Sb(0.428); ¹²⁴Sb(60 days), ¹²⁵Sb(2.7 yr); Sb₂S₃	100 mg day⁻¹ toxic to man	0.3	0.0033	1		2 × 10⁻⁴	2 × 10⁻³	0.0017	1·10⁻⁴	−0.2		3·10⁻⁴	−0.2	0.004	−0.2	
SELENIUM Se (AW 78.96)	⁸⁰Se (0.498) + ⁷⁴,⁷⁶,⁷⁷,⁷⁸,⁸²Se; With S ores; some amino acids, proteins, 1 plant wax	1–2 mg Se(IV) toxic to plants, 5 mg day⁻¹ toxic to man, 1–2 mg day⁻¹ lethal to rats	0.03	0.171	0.4		5 × 10⁻⁴	2 × 10⁻⁴			−63	0.0056	0.01–0.5	−30	0.42–1.9	0.17	Essential element for all vertebrates, protects against Ag, Cd, Hg and Tl
FLUORINE F (AW 18.9984)	¹⁹F(1); CaF₂, fluoro apatite; fluoro acids and nucleocidin	5 mg dm⁻³ toxic to plants; 20 mg day⁻¹ toxic and 2 g day⁻¹ lethal to man	220	0.5	200	1.4	0.1				1–400		3–19			0.05	1400	Accumulated by mammal bone and teeth

[a] In mg dm⁻³. [b] In ng m⁻³. [c] Concentrations quoted per unit dried weight. [d] Most of the Mn in sea water is in particulate form, as oxide fallout from volcanic dust.

Table 2 Listing of analytical problems in terms of the analyte, the complex matrix and the measurement technique

Analyte	Complex matrix	Measurement technique
Inorganic and organometallic analytes		
Na, K	Mineral water	Flame emission spectrometry
Ca, Mg	Water	EDTA titration
Ce	Ore	Gravimetry
Ti	Rock	Spectrophotometry
V(V)	Sea water	Adsorptive stripping voltammetry
Mo	River water, sea water	Solvent extraction/atomic absorption spectrometry
$Mn(PH_2O_2)H_2O$	Complex salt	Differential thermal analysis
Fe	Ore	Redox titration with $KMnO_4$
Co	Soil	Visible spectrophotometry
Ni	Steel	Gravimetry
Cu	Aqueous environment	Anodic stripping voltammetry
Organomercury compounds	Fish	High-performance liquid chromatography–cold vapour atomic absorption spectrometry
Zn	Pharmaceutical formulations	Ion-exchange separation and complexometric titration
B	Plant material	Visible spectrophotometry
Organolead compounds	Air	Gas chromatography–atomic absorption spectrometry
NO_2	Air	Sorbent tube collection–colorimetry
As	Hair	Neutron activation analysis
	Wallpaper	X-ray fluorescence spectrometry
Sb	Liver	Hydride generation–atomic absorption spectrometry
Se	Natural water	Spectrofluorimetry
F^-	Drinking water	Potentiometric measurement using an ion-selective electrode
Pb, Ba, Sn, I, Na, Ca, Mg, Fe, P, Cu, Zn, Rb, Br, Al, Mn	Biological fluids	Inductively coupled plasma mass spectrometry
Organic analytes		
Aspirin, phenacetin, caffeine, paracetamol, codeine, salicylamide	Pharmaceutical formulations	UV–visible spectrophotometry, NMR, voltammetry, high-performance liquid chromatography
Oxazepam and degradation products	Pharmaceutical formulations	High-performance liquid chromatography
Water-soluble vitamins	Formulations	Capillary zone electrophoresis and micellar electrophoretic capillary chromatography

Introduction 11

Table 2 (*continued*)

Analyte	Complex matrix	Measurement technique
Cephalosporins	Pharmaceutical formulations	Air-segmented continuous-flow visible spectrophotometry
Procaine penicillin G	Pharmaceutical raw material	Infrared spectrometry using dispersive and Fourier transform instruments
Polymer	Shoe-heel rubber	Thermogravimetric analysis
isocyanates	Workplace atmosphere	High-performance liquid chromatography with UV and electrochemical detection
Triazine pesticides	Environmental samples	Enzyme-linked immunosorbent assay
Ethyl acetate, higher alcohols	Beverage	Gas chromatography
Antibiotics	Dairy feedstuff	Gas chromatography–mass spectrometry
Cocaine and metabolites	Urine	Solid-phase extraction–gas chromatography
Explosives	Gunshot residue	High-performance liquid chromatography–electrochemical detection
Prozac and demethylated metabolite	Serum	High-performance liquid chromatography–fluorescence detection
Thioamides	Biological fluids	Cathodic stripping voltammetry
Haemoglobins	Biological fluids	High-performance liquid chromatography–electrospray mass spectrometry
Insulin	Biological fluids	Radioimmunoassay
Prostate-specific antigen	Human serum	Fluoroimmunoassay

Chapter 1
Historical Background

References to chemical analysis are to be found in the most ancient documents known to mankind. In the first century AD,[3] Plinius described the first spot test. Copper sulphate was often adulterated with iron sulphate and this was detected by spotting the test solution on a strip of papyrus soaked in the extract of gall-nuts. If iron was present, then it formed a complex with a component of the gall-nut extract, resulting in a black discoloration of the papyrus. Not surprisingly, little development took place in the Dark Ages, a time well described by the late Kenneth Clark in *Civilisation*: 'When life became too dangerous (in Europe) they struggled on in search of the most inaccessible fringes of Cornwall, Ireland or the Hebrides. They came in surprisingly large numbers. In the year 550 a boat-load of fifty scholars arrived at Cork. They wandered about the country looking for places that offered a modicum of security and a small group of like-minded men. And what places they found! Looking back from the great civilisations of twelfth-century France or seventeenth-century Rome, it is hard to believe that for quite a long time—almost a hundred years—western Christianity survived by clinging on to places like Skellig Michael, a pinnacle of rock eighteen miles from the Irish coast, rising seven-hundred feet out of the sea.'[4] The Middle Ages also passed by with alchemy being the predominant 'scientific' pursuit.

The rebirth or renaissance of the natural sciences had to await the arrival of the sixteenth century and particularly the application of inductive thinking (i.e. understanding phenomena in the natural sciences through experimentation) to the scientific method, as stressed by the English philosopher and statesman Francis Bacon (1561–1626). A great deal of work was also carried out at this time on the analysis of waters with respect to their medicinal effects. Libavius[3] in 1597 drew attention to the fact that the examination of mineral waters must take place near the spring so that the gaseous components (the *spiritus*) were not lost during transportation. He separated gaseous components from the aqueous sample, measured the quantity of dissolved salts by evaporation and identified some of these salts by crystal shapes grown on a thread (alum, vitriol and saltpetre). He used an extract of gall-nuts to identify iron and determined the ammonia content of the water by observing the blue colour formed when a copper solution was added.

Robert Boyle (1627–91), born at Lismore Castle, Co. Waterford, Ireland, was keen that chemistry should dissociate itself from the medical science to which it had become bound and it was he, in particular, that gave it new direction. Boyle is generally considered[3] to be not only the first British analytical chemist but also the father of analytical chemistry. He was the first to use the term 'chymical analysis,' to apply hydrogen sulphide for the identification of various metals and to define the sensitivity of chemical tests.[5] He is quoted, 'whence 'twas to conclude, that one grain of vitriolate substance ($FeSO_4$) would have impregnated six thousand times its weight of common water, so as to make it fit to produce with galls, a purple tincture.' He made a detailed study of precipitation and of the washing and drying of precipitates for gravimetric analysis and also contributed to gas analysis. He was also among the first to use acid–base indicators in the form of certain vegetable and animal extracts. Boyle returned to Ireland for only one extensive visit during his lifetime, to attend to the family estate in 1652–54. He had just recently started chemical experimentation[6] in England and was obviously very involved in his studies for he wrote in 1654 from Ireland to Frederick Clodius:[6,7] 'I live here in a barbarous country where chemical spirits are so misunderstood, and chemical instruments so unprocurable, that it is hard to have any hermetic thoughts in it, and impossible to bring them to experiment.... For my part, that I may not live wholly useless, or altogether a stranger in the study of Nature, since I wont for glasses and furnaces to make a chemical analysis of inanimate bodies, I am exercising myself in making anatomical dissections of living animals.'[8]

The 1700s saw significant developments in both inorganic and organic analysis in Britain and in Europe. For example, Francis Home in 1756 described a method which is one of the earliest examples of volumetric analysis. He assayed potash by measuring the number of teaspoons of dilute nitric acid required before effervescence ceased. William Lewis in 1767 was first to use a primary standard, a weight burette and a colour indicator.[9] Kirwan in 1784 used $K_4Fe(CN)_6$ as a standard solution for the determination of iron. Joseph Black of Glasgow and later of Edinburgh was the most outstanding teacher of analysis in the eighteenth century and he attracted many students from abroad. He differentiated magnesium and calcium carbonates by the shape of their crystalline salts and was first to use an indicator 'blank.'

Antoine Lament Lavoisier (1743–94) expounded the principles of combustion and indestructibility of matter in opposition to the phlogiston theory. This stated that all substances have a combustible component, phlogiston, and a non-combustible component so that when a substance is burnt, the phlogiston escapes and hence should cause a weight loss. In fact, as now appears so obvious, substances increase their weight on combustion. Lavoisier performed the first organic analytical experiments by combusting oils to produce carbon dioxide and water. The carbon dioxide was collected in sodium hydroxide and the water in anhydrous calcium chloride—semiquantitative results were obtained. Of the many books written about Lavoisier, some consider him the greatest chemist the world has known, while others are bitterly opposed to him and belittle him. It is possible that both are right, for he was a man of contrasts, being the greatest brain and most talented scientist of his age but with many human frailties. He was proud, vain and very desirous of glory and he was not ashamed to appropriate the discoveries of others and announce them as his own. He was guillotined in 1794 at the age of 51 for being a ruthless tax collector. As he boldly bowed his head under the guillotine, a witness who, it must be pointed out, was not one of his friends, declared, 'I do not know whether I saw the last and carefully played role of an actor, or whether my judgement of him before was wrong, and a really great man has died.'

Some of the important developments in inorganic analysis in the 1800s and early 1900s in Britain are given in Table 3, and it should be noted that the pattern of contributions in Europe and North America was similar to that in Britain.[10-12]

In organic analysis, Justus Liebig (1803–73) accurately analysed many organic substances for their carbon and hydrogen content and his method, based on the same principle as that used by Lavoisier, is still carried out today using his experimental conditions. He proved the old maxim that, 'without analysis there can be no synthesis,' or, paraphrased in a modern context, 'without analytical chemistry there can be no chemistry.' In contrast to many of the famous scientists who had been his predecessors, Liebig did not come from a rich and influential family. He did not show much ability during his early educational training and his teachers had no great hopes for his future. He consistently failed his examinations and was eventually apprenticed to a pharmacist. After much travelling, he was appointed Professor of Chemistry at the small University of Giessen at the age of 24. Many students

Table 3 Some important British contributions to analytical chemistry in the 1800s and early 1900s

1800	Cruickshank used electrolysis to detect copper
1803	Black observed the effect of CO_2 on titrations, noted indicator error and developed back-titration
1813	Use proposed 'Normal' solutions
1814	Wollaston's logarithmic tables
1816	Wollaston's 'analytical' slide-rule
1835	Reid—micro-scale teaching
1836	Marsh—test for arsenic
1841	Clark—hardness of water
1850	Penny—dichromate titration
1852	Andrews—microchemical methods
1852	Herepath—colorimetric method for iron
1871	Griess—nitrite test
1873	Hannay—electrolytic determination of mercury
1878	Bayley—spot tests on paper
1893	Thomson—glycerol–borate titration
1894	Fenton—dihydroxytartaric acid test for sodium
1903	Knecht and Hibbert—titanium(III) chloride as a titrant
1908	Sand—controlled potential, internal electrolysis
1915	Atack—alizarin as a colorimetric reagent for aluminium

attended the courses of this young and enthusiastic Professor, who sometimes had to teach in somewhat exacting conditions as illustrated by, 'I shall repeat the analysis more accurately as soon as the water in my laboratory thaws'! According to the evidence of his letters, Liebig's mood changed frequently; on some occasions he worked with feverish energy while on others he was depressed and his work suffered accordingly. To Wöhler, who synthesised urea and thus discredited the *vis vitalis* theory that organic substances were only produced by living organisms, he wrote, 'About my spirits I would not like to tell you. But I am almost sick of my life and often imagine that a shot or the cutting of my throat would help me in some cases. Oh, if only my unhappy life contained a few gay moments! If I did not have a wife and three children, I would be more content with a portion of hydrogen cyanide than with this life! To be truthful I hate chemistry and this terrible writing of books, which takes me to complete despair. I will write no more books, even if they would give me mountains of diamonds.' Wöhler attempted to reassure him: 'Dear friend, you suffer from the special disease of chemists, the so-called *hysteria chemicorum* which is caused by hard mental work, bad laboratory air and unlimited ambition. All great chemists suffer from this disease. You are wasting yourself and your health! Think of 1900, when we shall be once again carbon dioxide, water and ammonia—and what was once our bones will be part of the bones of a dog, which—who will worry then whether we had lived gaily or in bad temper? Who will know about your chemical debates? (He was a sharp and objective critic of many chemists, which made them enemies.) Nobody. The facts which you have disclosed, however, will always be known. But how do I attempt to advise the lion to eat sugar?' Away from his chemical pursuits, Liebig attempted writing popular books and founded a chemical fertiliser factory and plants for providing infant food, baking powder and coffee extract, all to no avail. He did, however, join a successful project to process meat from cattle slaughtered in Argentina for their hides. With time, Liebig's enthusiasm for teaching lapsed: 'Working with young men, which was a great joy before, is now a terrible job for me; a question or an explanation makes me almost miserable.' After a further Chair at the University of Munich failed to revive his research interests, he died in 1873, after writing to Wöhler a year before, 'I read scarcely any chemical literature. How can an old interest be extinguished to such an extent?'

These early elemental analyses were performed on macro samples and utilised gravimetric and titrimetric 'finishes.' This was adequate for organic analysis where grams of a substance were available, but the newer sciences of physiology, biology and biochemistry required microanalytical methods, since they were isolating much smaller amounts of substances in the milligram range. This scaling down of elemental analysis required the development of suitable sensitive microbalances, a task for which Fritz Pregl (1869–1930), working in Graz, was awarded the Nobel Prize in 1923, the first such award to be made for an achievement in analytical chemistry. With unlimited patience and great skill, working alone, he established the whole scheme of quantitative microanalysis. He began with carbon and hydrogen determinations and continued his work with the determination of nitrogen, halogens, sulphur, methoxy and ester functions. In most cases, the principle of the method was devised by Pregl, but in all cases, the apparatus, instrumentation and method itself were the work of Pregl alone. In 1917, he wrote all his work up in a book, *Die Quantitative Organische Mikroanalyse*, which opened up a new field of analytical chemistry.

Such quantitative organic analysis was obviously developed before the corresponding qualitative analysis. When the difficulties inherent in qualitative organic analysis are considered, then this becomes understandable. In inorganic analysis, a qualitative knowledge of the elemental composition is often sufficient to identify the compound (for example, calcium carbonate, $CaCO_3$), whereas for an organic compound such as glucose, $C_6H_{12}O_6$ (**I**), even a quantitative elemental analysis can give little information as to its structure. Ref. 13 gives a total of 27 possible structures for $C_6H_{12}O_6$, including monosaccharides such as glucose (**I**), inositol (**II**) and psicose (pseudofructose, **III**).

Modern qualitative organic analysis, dealing with a vast number of recorded compounds, is thus unthinkable without the developments in analytical instrumentation in the twentieth century (e.g. infrared spectrometry, nuclear magnetic resonance spectrometry, mass spectrometry). Other instrumental developments such as atomic absorption spectrometry and inductively coupled plasma mass spectrometry have been of great value in the development of qualitative inorganic analysis.

Analytical chemistry has thus traditionally been interwoven with inorganic and organic chemistry and, in more recent times, organometallic chemistry and interdisciplinary subjects involving chemistry for research into and development of methods for the qualitative and quantitative analysis of many chemical compounds. Many of the instrumental techniques employed in physical chemistry for investigating the properties of atoms and molecules have also been used in analytical chemistry both for the elucidation of chemical constitution and structure and also for quantitative analysis. For quantitative analysis, a mathematical relationship must exist between the *event*, such as a physical process like the absorption of radiation or the transfer of electrons by an atom or molecule, *converted into an electrical signal* (process of transduction) and the *concentration* of the atoms or molecules. Reference can be made to a relevant statement by the eighteenth century German philosopher, Immanuel Kant: 'In the various branches of the natural sciences only that which can be expressed by mathematics is true science.'

It is not difficult to see that, at certain times and in certain establishments, analytical chemistry would be seen as a less-than-equal member of the overall team since it had to provide a service facility to verify the chemical constitution and structure of the products of creative synthetic chemistry, using many of the tools designed in physical chemistry. A parallelism can be observed in the chemical analysis of an object of art such as a painting, in an attempt to detect a forgery. Here, creativity still belongs to the artist, whoever he or she may be, and it is for the techniques of analytical chemistry such as infrared reflectography to be used to observe this work of art through another eye, to establish its authenticity or otherwise. The analytical chemist or scientist must therefore possess and develop an art of his or her own in the exercise of his or her human skill in order to obtain accurate, precise and meaningful data on the samples provided. This 'art' of the analytical scientist is particularly in evidence and demand in interdisciplinary circles when the identification and quantitative determination of atoms, ions and molecules in the complex samples of today's world is reviewed, i.e. concerning samples taken from the atmosphere and aquatic environments, industrial products, drugs, foods, body fluids, plants, soils, etc.

Through unravelling the complexities of chemical constitution, structure and concentration in many of these samples, the analytical scientist can only but wonder at the beauty of the chemical creation itself. It is a search for such wonder, not to speak of the untold benefits that these investigations have for man's overall well-being on this planet, that the author feels will spur on those scientists, teachers, students and lay people alike in the many and varied chemical measurements carried out on themselves, their world and their meaning.

Chapter 2

Unit Processes of Analytical Procedures

Textbooks in analytical chemistry are generally organised according to the subdivisions of analytical techniques, whether they be chemical or instrumental, e.g. gravimetric and volumetric analysis, electrochemical techniques of analysis such as potentiometry, polarography and coulometry, analytical atomic and molecular spectroscopy, separation techniques such as solvent extraction, thin-layer, gas and liquid chromatography, thermal, radiochemical and biological methods of analysis. While it is appreciated that the students must first understand the theory and applications of these individual analytical techniques, there is a dearth of literature on how to go about solving the many and varied analytical problems that are encountered in today's world. These problems, which essentially involve the identification and/or quantitative determination of a wide variety of atoms, ions and molecules in samples such as air, water, soil, food and body fluids, can be subdivided into their constituent unit processes: (i) definition of the problem, (ii) choice of the method/technique, (iii) obtaining a representative sample and its measurement, (iv) preliminary treatment of the sample, (v) separation of analyte(s) from interferences and from each other, (vi) measurement, (vii) statistical assessment of measurements and (viii) calculation of analytical results and solution of the problem. The analytical scientist, in the solution of such a problem, has to be able to design and/or carry out an appropriate analytical method with the measurements on the complex sample being interpreted within the context of the original problem. This generally requires a wide knowledge of chemistry, aspects of biochemistry, law, physics, economics, familiarity with and ability to operate a varied range of modern analytical instruments and an understanding of the meaning of the measurements obtained. It has also to be stressed that the measurement step, i.e. the 'finish,' frequently performed these days by sophisticated microprocessor-controlled instruments, is usually carried out on a solution freed from interferences and is only one unit process in the overall analytical method. When other unit processes such as sampling, preliminary treatment of sample and separation are poorly considered, the relationship of the analytical result to the population from which the sample was taken is uncertain and impossible to interpret.

2.1 UNIT PROCESS NO. 1. DEFINITION OF THE PROBLEM

An analytical problem can be defined as the identification and/or determination of a chemical species such as an atom, ion or molecule in a particular complex matrix. Such a species may be inorganic, organic or organometallic in nature. Sample sizes can be classified[14] as gram (1–10 g) through to milligram (0.001–0.01 g), microgram (10^{-6}–10^{-3} g) and down to femtogram (10^{-15}–10^{-12} g). Macro samples are considered as weighing 0.1–1.0 g, micro samples as weighing 10^{-2}–10^{-3} g (hence the term microanalysis) down to ultramicro samples which weigh $< 10^{-4}$ g. Species in a sample may exist as major constituents (100–1%), minor constituents (1–0.01%), traces (10^{-2}–10^{-8}%, which is 10^{2}–10^{-4} ppm), down to picotraces (10^{-10}–10^{-13} ppm).[14] Most of the problems dealt with in this text will have a sample size of 1–100 g and constituent species in the sample will be at major, minor or trace concentrations. Note that 1 ppm or 10^{-4}% refers to 1 part of the trace species in 1 million parts of sample [e.g. 1 µg species (cm sample)$^{-3}$, 1 µg species (g sample)$^{-1}$] and that 1 ppb or 10^{-7}% refers to 1 part of the trace species in 10^{9} parts of sample (US billion) [e.g., 1 ng species (cm sample)$^{-3}$ or 1 ng species (g sample)$^{-1}$]. In major constituent determinations as with some pharmaceutical formulations, concentration can be expressed in parts per hundred, i.e. % (w/w, w/v, v/v).

Examples of topical analytical problems which illustrate the importance of the problem solving approach of *Analytical Chemistry of Complex Matrices* are given in Table 4. These are in the form of questions on 'what do we really want to know,' answered by the application of an analytical method to a particular complex matrix which, more often than not these days, embraces a modern instrumental technique of analysis. Some of these analytical problems are also discussed at length in this volume, e.g. inductively coupled plasma mass spectrometry for the rapid identification of trace elements in a particular sample (Section 4.11), the determination of arsenic in Napoleon's hair (Section 4.7) and assay of an iron ore (Section 3.8). The overall contents, however, of *Analytical Chemistry of Complex Matrices*, some of which are of particular current interest, are given in Table 2 and were primarily chosen so as to give a representative selection of inorganic, organic and organometallic analytes to be identified and/or quantitatively determined in a variety of complex matrices, using a range of classical chemical and instrumental techniques of analysis. The author pays particular attention to the underlying chemistry of each analytical method and the wide selection of analytes, matrices and techniques chosen so as to be of value to both undergraduate and postgraduate scientists involved in qualitative and quantitative analysis of complex matrices and to analytical scientists in government and industrial laboratories who are inexperienced with particular measurements.

2.2 UNIT PROCESS NO. 2. CHOICE OF METHOD

The choice of the method is dependent on the analytical methodology and instrumentation in the analytical laboratory that has to solve the problem. A literature

Table 4 Illustration of the importance of the problem-solving approach of *Analytical Chemistry of Complex Matrices* in modern life

Problem—what do we really want to know?	Analytical method	Solution to problem	Ref.
How would you assess the intake of human carcinogens such as aromatic amines in passive smoking situations?	Cigarettes smoked in a home-made smoking apparatus. Samples of mainstream and sidestream smoke were trapped in Drechsel bottles containing 5% HCl and three internal standards prior to solvent extraction, derivatisation and gas chromatography–mass spectrometry. The analytical methodology was then applied to air samples taken from selected passive smoking situations	Aromatic amines are present in the air of rooms where people smoke (barber shops, movie theatres, etc.) In large-space enclosures such as trains and aircraft detectable levels of aromatic amines can be found in no-smoking areas. Prolonged ventilation reduces the levels of aromatic amines only to half the original concentration. Concentrations in these studies were reported as total aromatic amines in ng m^{-3}	15
How would you maintain maximum tolerable blood levels of the anticancer drug dacarbazine in a patient undergoing chemotherapy over a five-day period?	Sample blood regularly and subject it to separation procedures followed by high-performance liquid chromatography–UV detection	Dacarbazine elutes separately from biological interferences at ca 14 min, permitting its quantitation and hence providing the clinician with data on which to base administration of the drug	16
The veterinary pharmaceutical levamisole can have a structurally related toxic impurity present, i.e. 2,3-dihydro-6-phenylimidazo-(2,1,*b*)tiiazole. How would you maintain levels below a certain threshold value by a simple, inexpensive and rapid test?	Use thin-layer chromatography on silica gel with toluene–methanol–glacial acetic acid $(90 + 16 + 8)$ as the mobile phase, visualising with potassium iodoplatinate solution. Also chromatograph a British Pharmacopoeia Chemical Reference Standard (BPCRS) made up to 0.025% w/v	10 µl of 0.025% w/v BPCRS is compared with 10 µl of a 5% w/v solution of the suspect raw material. The TLC spot with the latter solution should *not* be more intense than the spot with the former solution	17

			21
A firm proposes to purchase iron ore from a new source. How would you accurately assay a representative sample for its Fe content, bearing in mind that a small difference in the assay figure can be of commercial significance?	Dissolve the sample and determine % Fe by redox titration as discussed in Section 3.8	A well trained analytical scientist can achieve highly accurate results by this inexpensive, albeit time-consuming, classical method	Section 3.8
Furthermore, it is believed that this ore also contains trace amounts of certain valuable metals. How would you rapidly identify them?	Aspirate solubilised sample into inductively coupled plasma mass spectrometric apparatus	A rapid identification of the nature of the trace elements in the ore sample can be made, albeit using expensive analytical instrumentation which is unlikely to be available in the firm's laboratory. Sample should be sent to an appropriate university or industrial laboratory	Section 4.11
Triphenyltin hydroxide has been applied as a fungicide to rice-field water. How would you ascertain the concentration of the active Ph_3Sn^+ in the water is maintained at an optimum level?	Without resort to an expensive coupled technique such as high-performance liquid chromatography–atomic absorption spectrometry, simply record the differential-pulse polarogram in pH 7 buffer with 50% ethanol	At trace concentrations of Ph_3Sn^+, a well defined polarographic peak is observed corresponding to the adsorption prewave of the process $Ph_3Sn^+ + e \rightarrow Ph_3Sn^{\cdot}$. The method depends on there being no significant interferences in the neighbourhood of the appropriate voltammetric signal, and can be calibrated using standard Ph_3SnOH solutions	18

(continued overleaf)

Table 4 (continued)

Problem—what do we really want to know?	Analytical method	Solution to problem	Ref.
Was Napoleon in exile subjected to deliberate arsenic poisoning?	Using preserved hair samples, subject these to neutron activation analysis	Napoleon's hair did not contain high levels of arsenic and it is probable that the arsenic in his hair came from volatile $As(CH_3)_3$ produced metabolically by moulds growing on As-loaded wallpaper his rooms. It is therefore unlikely that Napoleon was deliberately poisoned with arsenic	Section 4.7
Isoprene is a major hydrocarbon emission from biogenic sources such as a eucalyptus forest. Quantify these emissions *in situ*	Ambient air samples taken using a personal sampling pump and a cooled sample adsorption tube at about 16 m height within the forest canopy. Immediately transfer the tube to a programmed-temperature vaporisation unit for thermal desorption of analytes to a wide-bore thick-film capillary GC column	Isoprene and its oxidation products (methacrolein, methyl vinyl ketone) measured by flame ionisation detection following gas chromatography in an on-site mobile laboratory. Diurnal variations in the concentrations of these molecules were clearly visible. Field concentrations of isoprene are of the order of 1–20 ppb.	19
Sedimentary rock cores of age 65 M years have been removed from floor of the Atlantic Ocean near South Africa. If the iridium concentrations suggest that the extinction of the dinosaurs was due to the collision of iridium-rich 10-km diameter asteroids with the earth at that time, how would you determine the iridium concentration?	Dissolve the sample appropriately and subject it to atomic absorption spectrometry with a flame and/or graphite furnace, using an iridium hollow-cathode lamp	Iridium concentrations in the 65M-year-old strata are found to be significantly higher than iridium in other core samples. The hypothesis would appear to have some validity	

| Has the suspect in the bombing handled nitroglycerine or not? | Swab hands of suspect with diethyl ether. Carry out Griess test by reaction of residue with 0.1% NaOH followed by reaction with sulphanilic acid and α-naphthylamine at 4–6°C | If nitroglycerine is present, a pink colour should be generated within 10 s. However, the concentration of NaOH used to generate NO_2^- prior to formation of a coloured azo dye is critical. If a 1% solution of NaOH is used, then nitrocellulose from cigarette packets also gives this reaction. It is therefore essential that an alternative analytical method such as TLC be applied to confirm the presence or otherwise of nitroglycerine in such a sensitive situation |

search is advisable at this stage so that the analytical scientist can acquaint himself or herself with alternative analytical methods of which he or she is not aware or may have forgotten. Even existing methods may need major improvements to suit the requirements of the laboratory that will perform the measurements to solve the problem. The choice of the method is also dependent on whether the chosen analytical method will give accurate and precise results within a given concentration range with an acceptable limit of detection and, in addition, give results with the necessary selectivity, time per sample, cost to client, etc. The analytical scientist should first of all give the chosen analytical method several trial runs in his or her laboratory to ensure the method's reliability and conformance with the above requirements. It should be noted that relatively simply analytical methods can be automated and this is of significant value in the handling of large numbers of samples, say in clinical biochemistry.

It is an increasing requirement that the overall analytical method should be suitable for its intended purpose. This process must verify that all the steps needed to generate qualitative and quantitative analytical results are valid, as illustrated in Figure 2 with high-performance liquid chromatography (HPLC) and capillary electrophoresis (CE) in mind. Such validation is particularly important in commercial life, where confidence in measurement plays a vital part in sustaining industrial competitiveness. It is estimated that chemical analysis is an activity that engages about 200 000 people in the UK at an annual cost of approximately £4 billion. The UK Laboratory of the Government Chemist believes that as many as 20% of these analyses are not fit for their intended purposes. Application of Valid Analytical Measurement (VAM) can therefore minimise disputed results, repeated analyses and all the extra costs of contention. The four key principles of VAM are:

(i) measurements should be made using properly validated methods, as already discussed;
(ii) quality assurance protocols should incorporate certified reference materials to ensure traceability of measurements;
(iii) laboratories should seek an independent assessment of their performance for particular analyses, preferably by participating in national and international proficiency testing schemes. This is further discussed later in this section.
(iv) laboratories should seek independent approval of their quality assurance arrangements, preferably by accreditation or licensing to a recognised quality standard such as that supplied by NAMAS (National Analytical Measurement Accreditation Service), GLP (Good Laboratory Practice, e.g. DOH Guidelines, 1989) or ISO 9000 (International Standards Organisation).

At this stage, it is worthwhile defining parameters important in the choice of method.

ACCURACY

This is the nearness of a measurement to the true value. It is expressed in terms of error, which is the difference between the true result and the measured value (abs-

Unit Processes of Analytical Procedures 25

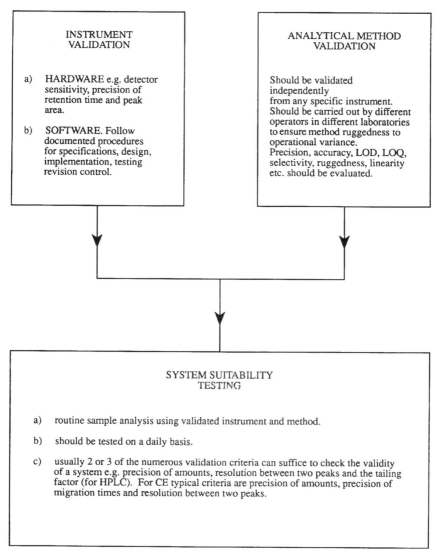

Figure 2 Validation steps necessary to verify that an overall analytical method is suitable for its intended purpose

olute error). Relative error is the error expressed as a percentage of the measured value. Accuracy can be determined by replicate analysis of a sample containing a known amount of the analyte. Usually three concentrations (low, intermediate and high) are tested to ascertain whether accuracy is concentration dependent.

The accuracy of a measurement has achieved particular prominence of late in attempting to harmonise results produced in different laboratories in different countries on the 'same' sample. The recently published *International Harmonised Protocol on Proficiency Testing*[20] has involved a collaboration of scientists from many countries under the joint organisation of ISO, IUPAC and AOAC International. On the technical side, aspects of proficiency testing given prominence in the Protocol are the establishment of the stability and effective homogeneity of the materials to be circulated to the participating laboratories. Most analytical proficiency tests involve the distribution to each participant of a sample taken from a homogeneous bulk material. This is then subjected to the participating laboratory's analytical methodology for measurement of the particular analyte in the particular matrix. The result (x) is converted into a 'z score' according to

$$z = \frac{x - \hat{x}}{\sigma} \tag{1}$$

where $\hat{x}$ is the assigned value and σ is the 'target value for standard deviation.' The latter parameter describes the acceptable variability between the laboratories and is related to fitness of the analysis for its intended purpose. The assigned value, $\hat{x}$, is a genuine estimate of the true value and is arrived at by consensus of expert laboratories using recognised methods under good conditions.

z Scores of between $+2$ and -2 are regarded as satisfactory but those outside the range $+3$ to -3 as unsatisfactory. Once the individual z scores have been submitted, then they can be presented as in Figure 3, which is a typical chart from FAPAS (Food Analysis Performance Assessment Scheme). Each participating laboratory can then see their result, x, in relation to the assigned value $\hat{x}$ and the target value of standard deviation and compare it with those x values obtained by the other participating laboratories. It is believed that satisfactory results, i.e $|z| \leqslant 2$, will generally occur in about 95% of cases, questionable results, i.e. $2 < |z| < 3$, in about 5% of cases and unsatisfactory results, i.e. $|z| \geqslant 3$, in about 0.3% of cases.[21]

PRECISION

This is the variability of a measurement and is best expressed as the standard deviation, S:

$$S = \left[\frac{\sum_{i=1}^{i=N}(x_i - \bar{x})^2}{N - 1} \right]^{1/2} \tag{2}$$

where N is the number of results, x_i is the measured result and $\bar{x}$ is the mean of the set of results. A better estimate of the standard deviation may often be obtained by the pooling of results from more than one set. Thus S can be calculated from K sets of data:

$$S = \left[\frac{\sum_{i=1}^{N_1}(x_i - \bar{x}_1)^2 + \sum_{i=1}^{N_2}(x_i - \bar{x}_2)^2 + \cdots + \sum_{i=1}^{N_K}(x_i - \bar{x}_K)^2}{M - K} \right]^{1/2} \tag{3}$$

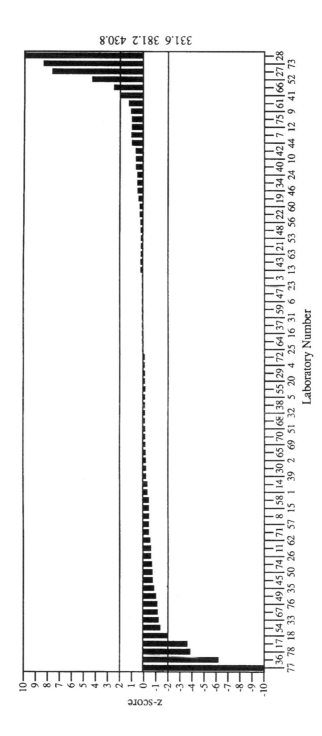

Figure 3 FAPAS (Ford Analysis Performance Assessment Scheme) Report 0307 on benzoic acid (381.2 mg dm^{-3}) in Orange Drink I. Reproduced by permission of the Laboratory of Government Chemist, London

where $M = N_1 + N_2 + \ldots + N_K$. One degree of freedom is lost with each set pooled. The *relative standard deviation* or *coefficient of variation* $[(S/\bar{x}) \times 100]$ is useful in comparing precisions.

The within-day precision, in a field such as therapeutic drug monitoring, can be determined by duplicate measurements of drug levels in at least 10 samples for a particular concentration. It is recommended that this is repeated for two other concentrations so that the linear calibration range for expected drug concentrations is covered. For day-to-day measurement of precision, each sample representing a particular concentration should be stored and treated in exactly the same way prior to determination. If, for example, lyophilised serum is used, the same recommended reconstitution procedure should be used each time.

In 1987, Horwitz and Albert[22] illustrated how the coefficient of variation for topical analyses carried out in different laboratories increased with decrease in concentration. Significantly large coefficients of variation were observed between laboratories for the determination of ppm concentrations of trace elements and pesticide residues, ppb concentrations of aflatoxins and sub-ppb concentrations of dioxins. This is illustrated in Figure 4, otherwise known as the 'Horwitz trumpet.' The variation of the coefficient of variation with concentration in the determination of aflatoxins in peanuts is further discussed in Section 2.3.1 to illustrate the importance of sampling in this analytical method.

SENSITIVITY

The sensitivity of an instrumental technique can be confused in the literature with its limit of detection. In atomic absorption spectrometry, sensitivity is defined as the concentration of an element in ppm which produces a transmittance signal of 0.99 or a corresponding absorbance signal of 0.0044, whereas the limit of detection is defined as the concentration of the element that produces an analytical signal equal to twice the standard deviation of the background signal. For flame atomisation, the standard deviation of the background signal is found by observing the signal variation while a blank is aspirated into the flame. Both the sensitivity and the limit of detection are affected by variables such as temperature, spectral bandwidth, detector sensitivity and type of signal processing. Small differences such as a factor of 2–3 among quoted values of the two parameters are not significant, but an order of magnitude difference certainly would be significant. Table 5 lists some sensitivities and limits of detection for selected elements using various atomic spectrometric methods.

In general terms, sensitivity can be defined as the slope of the calibration plot, i.e. tan α, where α is the angle between the calibration plot and the concentration or x axis. Sensitivity is therefore the change in signal per unit of concentration and, as such, has different units depending on the instrumental technique to which it relates.

LIMIT OF DETECTION (LOD) AND LIMIT OF QUANTITATION (LOQ)

The limit of detection is the lowest concentration of an analyte that an analytical system can reliably detect. It is a statistical definition, related to the random errors of

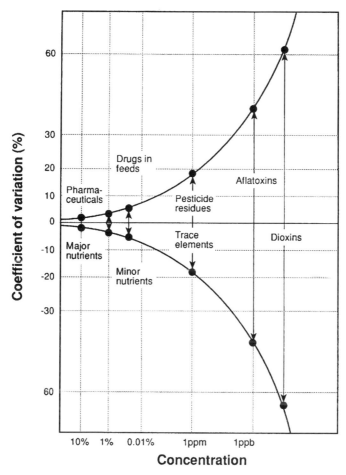

Figure 4 Dependence of coefficient of variation on concentration for topical analyses. Reproduced by permission of the Royal Society of Chemistry from W. Horwitz and R. Albert, *Anal. Proc.*, 1987, **24**, 49

an analytical measurement. The term 'analytical system' is used instead of 'analytical method' since 'the method' *per se* constitutes only part of the total variation observed during its use. Such additional factors include the analyst, the environment, the brand of instrument used, the quality of reagents, the nature of samples, the protocols used for calibration and reagent blank correction. It is these additional factors that, together with the method, constitute the analytical system. Therefore, in quoting a particular detection limit, details of the analytical system, to which the LOD applies, should be given in the text.

The LOD of most analytical systems involving an instrumental technique is based on the relationship between the gross analyte signal S_t (i.e. signal of analyte and

Table 5 Sensitivity and limits of detection for selected elements by various atomic spectrometric methods

Element	AA (flame) sensitivity[23] (ppb)	AA (electrothermal) sensitivity[23] (ppb)	AE (ICP) limit of detection[23] (ppb)	AE (flame) limit of detection[24] (ppb)
As	100	0.2	2	—
Ba	200	0.06	0.02	1.0
Ca	20	0.4	0.001	0.1
Cd	10	0.003	0.06	2000
Co	70	0.06	0.1	50
Cu	40	0.04	0.4	10
Fe	60	0.008	0.5	50
Hg	2200	0.5	1.0	—
K	10	0.1	—	0.5
Na	3	0.004	0.2	0.5
Pb	100	0.05	1.0	200
Zn	9	0.002	0.1	—

blank), the blank signal S_b and the standard deviation of the blank signal, σ_b. The LOD has been defined by the extent to which the gross signal exceeds S_b:

$$S_t - S_b = k_d \sigma_b \tag{4}$$

If a single sample is being analysed for which there are no blank data, then the LOD is based on peak-to-peak noise ($\sigma_b = \sigma_n$) measured on the baseline close to the actual or expected analyte peak. For many years a value of $k_d = 2$ has been used, although in recent years a minimum value of $k_d = 3$ has been recommended.[25–27] Thus the LOD is located at $3\sigma_b$ above the blank signal S_b. Using this definition of LOD, the normal distribution curves of the blank signal S_b and the gross signal S_t will overlap to the extent that the probability of a false positive (concluding the analyte is present when it is absent) and a false negative (the reverse) is 7%. Limit of detection in signal units can then be converted into limit of detection in concentration units using the calibration plot.

It is also possible to obtain the terms S_b and σ_b and hence LOD using a conventional regression line obtained from the method of least squares. It is assumed that the standard deviation about the regression line $\sigma_{y/x}$ is equal to σ_b and that the value of the calculated intercept on the signal axis is used as an estimate for S_b. Hence the LOD can be calculated in signal units and converted into concentration units using the regression equation.

> For example, standard aqueous solutions of fluorescein are examined in a fluorescence spectrometer and yield the fluorescence intensities (in arbitrary units) shown in Table 6.
> The regression equation can be calculated as $y = 1.93x + 1.52$, i.e. intercept on signal axis $= 1.52$ and slope $= 1.93$ with correlation coefficient $r = 0.9989$ and standard deviation about the regression line $\sigma_{y/x} = 0.4329$. The value of the signal at the LOD is therefore $1.52 + (3 \times 0.4329) = 2.82$. Use of the regression equation then yields an LOD of 0.67 pg cm^{-3}.

Unit Processes of Analytical Procedures

Table 6 Fluorescence intensities (in arbitrary units) observed for pg cm^{-3} concentrations of fluorescein

Fluorescence intensity (arbitrary units)	Fluorescein concentration (pg cm^{-3})
2.1	0
5.0	2
9.0	4
12.6	6
17.3	8
21.0	10
24.7	12

The LOD can also be calculated using a statistical equation derived from comparison of two experimental means:[41]

$$\text{LOD} = t\sigma_b \sqrt{\frac{N + N_b}{NN_b}} \qquad (5)$$

where N_b is the number of blank determinations and N is the number of analyses carried out for the analyte.

For example, a method for the analysis of DDT gave the following results when applied to pesticide-free foliage samples: DDT = 0.2, −0.5, −0.2, 1.0, 0.8, −0.6, 0.4, 1.3 µg. Calculate the LOD at the 99% confidence level of the method for a single analysis.

The mean blank value is 0.3 µg and the σ_b value can be calculated as 0.68 µg. Therefore, for a single analysis $N = 1$ and the number of degrees of freedom will be $(1 + 8 - 2) = 7$. From t tables (Table 13), $t = 3.5$ and therefore

$$\text{LOD} = 3.5 \times 0.68 \sqrt{\frac{1 + 8}{1 \times 8}} = 2.5 \; \mu\text{g DDT}$$

Thus 99 times out of 100, a result of 2.5 µg DDT will indicate the presence of DDT in the plant.

The limit of quantitation is located above the measured average blank S_b by the following definition:

$$S_t - S_b = k_q \sigma_b \qquad (6)$$

It is recommended that the minimum value of k_q be 10. Thus the LOQ is located $10\sigma_b$ above the blank signal S_b. LOQ in signal units can then be converted into LOQ in concentration units using the calibration plot. The LOQ is generally regarded as the lowest concentration at which reliable and precise calibration plots may commence.

SELECTIVITY

Selectivity, in general terms, is the ability of the analytical method as a whole, or technique in particular, to discriminate between the analyte and potential interferences. The selectivity of an analytical method can be defined as the ratio of the interferent to analyte which can be tolerated to the stated precision on application of the whole analytical method, e.g. 100 μg Sb in the assay of 10 μg As in, say, 1 g GeO_2 at $\pm 2\%$ relative standard deviation gives a 10-fold overall selectivity. Instrumental selectivity, in terms of the calibration plot of response vs concentration, can be defined as XZ/YZ for a particular concentration, where larger values of this ratio refer to greater selectivity. Figure 5 illustrates this for the atomic absorption spectrometric determination of Ca in the presence of a spectral interference.

In chromatography, selectivity can be understood in terms of the resolution R,

$$R = \frac{2\Delta t_R}{W_A + W_B} \tag{7}$$

where Δt_R is the difference in retention times of peaks A and B and W_A and W_B are the base widths of peaks A and B in time units identical with those for Δt_R. When $R \geq 1.5$, optimum chromatographic resolution or selectivity is said to be obtained, since the overlap of peaks A and B is $\leq 0.1\%$. In those chromatographic situations where only a single peak is obtained in the time scale of the separation, there is the possibility that the separation may have a high selectivity (i.e. high peak purity) or a low selectivity with impurities co-eluting with the particular analyte peak. Peak purity analysis (PKPA) software is available[28] for use on either HPLC or CE signals

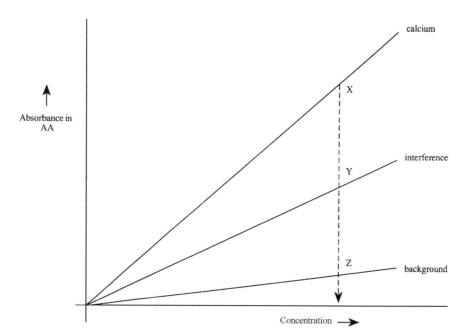

Figure 5 Definition of instrumental selectivity in terms of the calibration plot of absorbance vs concentration for the atomic absorption determination of Ca in the presence of a spectral interference

or using spectral data generated by a diode-array detector (DAD). When a co-eluting impurity is present at a significant concentration, a spectral overlay of eluting peak and standard analyte, supported by data treatment such as difference spectra, may be adequate for impurity detection. As a rule of thumb, impurity concentrations in the 0.1–1% range may be detected when the spectra are dissimilar, whereas the impurity detection limit is of the order of 5% when the spectra are similar. Statistical software routines can be used as the concentration of the impurity diminishes further and its impact on the analyte spectrum becomes more subtle. The similarity factor is a numerical index of the 'goodness of fit' between two spectra. It is given by the least-squares fit coefficients computed for all absorbance pairs in the spectra under comparison, i.e.

$$\text{Similarity factor} = 1000 \left\{ \frac{\sum_{i=1}^{i=n}[(A_i - A_{av})(B_i - B_{av})]}{\sqrt{\left[\sum_{i=1}^{i=n}(A_i - A_{av})^2\right]\left[\sum_{i=1}^{i=n}(B_i - B_{av})^2\right]}} \right\}^2 \quad (8)$$

where A_i and B_i are the respective absorbances measured at the same wavelength in the first and second spectra, n is the number of data points and A_{av} and B_{av} are the average absorbances of the first and second spectra, respectively. A similarity factor of 1000 indicates identical spectra. A factor close to an ideal match (> 990) is generally considered to result from identical spectra. Similarity factors in the 900–990 range indicate some similarity but require more careful consideration and interpretation of the comparative data.

Figure 6 shows the reversed-phase HPLC separation of six reactive textile dyes such as Remazol Black B (**IV**) and some process impurities/degradation products using a mobile phase of acetonitrile–water $(60 + 40)$ containing 4.5×10^{-4} mol dm^{-3} cetyltrimethylammonium bromide (CTAB).[29] CTAB is an ion-pairing agent which ion pairs to anionic groups such as the sulphonates in **IV**, thereby increasing their hydrophobicity and retention time on the C_{18} column, thus allowing for a more favourable separation. Baseline resolution (i.e. $R > 1.5$) is achieved for three of these molecules (peaks 2, 3 and 4). An R value of 0.88 was obtained for the separation of Remazol Red RB (peak 9) and co-eluting Remazol Black B and Remazol Navy Blue GG (peak 7, 8). Cibacron Orange CG (peak 5) essentially co-elutes with a Remazol Black B impurity (peak 6). PKPA software would, for example, confirm co-elution of Remazol Black B and Remazol Navy Blue GG. Significantly improved separation can be achieved by recourse to micellar electrokinetic capillary chromatography of the hydrolysed dyes (Figure 7), which exhibits complete baseline resolution (i.e. $R > 1.5$ in all cases) of the six textile dyes and some process impurities/degradation products.[29]

The selectivity coefficient, K, of an ion-exchange resin, Res$^-$, represented by

$$Na^+Res^- + K^+(aq) \rightleftharpoons K^+Res^- + Na^+(aq)$$

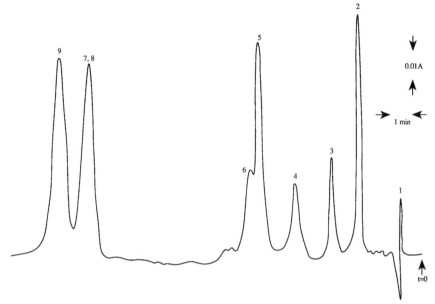

Figure 6 Analysis of reactive dyes by reversed-phase HPLC. (1) Solvent front; (2) Remazol Golden Yellow RNL; (3) Cibacron Red C-2G; (4) Remazol Black B impurity; (5) Cibacron Orange C-G; (6) Remazol Black B impurity; (7 and 8) Remazol Black B, Remazol Navy Blue GG; (9) Remazol Red RB. Conditions: mobile phase, acetonitrile–water (60 + 40) containing 4.5×10^{-4} mol dm^{-3} CTAB; column, 3.9 mm i.d. × 150 mm Nova-Pak C$_{18}$; flow rate, 0.7 cm^3 min^{-1}; injection, 20 µl of a 10^{-5} mol dm^{-3} solution of dyes (2), (3) and (5); other dyes at a concentration of 10^{-4} mol dm^{-3}; detection, visible absorbance at 485 nm. Reproduced by permission of John Wiley & Sons Ltd from D. A. Oxspring, E. O'Kane, R. Marchant and W. F. Smyth, *Anal. Methods Instrum.*, 1994, 1(4), 19

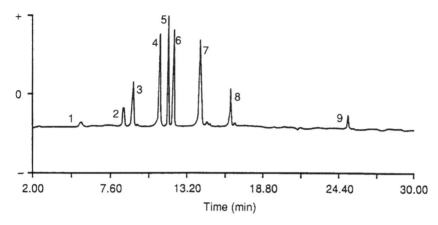

Figure 7 Separation of hydrolysed reactive dyes. (1) Acetonitrile front (5.53 min); (2) Remazol Golden Yellow RNL impurity (8.53 min); (3) Remazol Golden Yellow RNL (9.25 min); (4) Cibacron Red C-2G (11.20 min); (5) Remazol Navy Blue GG (11.84 min); (6) Remazol Black B (12.26 min); (7) Cibacron Orange CG (14.17 min); (8) Remazol Red RB (16.40 min); (9) Cibacron Orange CG impurity (24.9 min). Reproduced by permission of John Wiley & Sons Ltd from D. A. Oxspring, E. O'Kane, R. Marchant and W. F. Smyth, *Anal. Methods Instrum.*, 1994, **1**(4), 19

is given by

$$K = \frac{[K^+(Res)][Na^+(aq)]}{[Na^+(Res)][K^+(aq)]} \qquad (9)$$

where $[K^+(Res)]$ and $[Na^+(Res)]$ are the respective concentrations of K^+ and Na^+ bound to the ion-exchange sites on the resin. K is then a measure of the relative selectivity of the resin for K^+ over Na^+.

In potentiometry, for the measurement of high pH values of > 12 by a glass electrode, the selectivity of the electrode for H_3O^+ over Na^+ is given by the selectivity constant $K_{H_3O^+,Na^+}$ in the expanded Nernst equation:

$$E = E_k + 0.0591 \log (a_{H_3O^+} + K_{H_3O^+,Na^+} \cdot a_{Na^+}) \qquad (10)$$

Many published analytical methods express selectivity in terms of the effects of a stated concentration of an interference on the recovery of the analyte. For example, microgram quantities of Se can be determined by reduction of Se in samples taken from the environment to Se^{2-}, acidification to produce H_2Se, which is then trapped in an alkaline solution to be determined by cathodic stripping voltammetry (CSV) as HgSe. Sulphur is a major interferent since S^{2-} will plate on to the hanging mercury drop electrode as HgS and then reduce at a similar potential as HgSe (at pH 12, the stripping peak potential of HgS is $E_p = -0.78$ V and that of HgSe is $E_p = -0.90$ V). Metal ions such as Cu^{2+}, Ni^{2+}, Co^{2+}, Zn^{2+} and Pb^{2+} also seriously interfere owing to the formation of insoluble metal selenides. These effects are illustrated in Table 7.[30]

Table 7 Interferences in the CSV determination of selenium

Species taken	Amount added (μg)	Recovery of Se (%)
SiO_3^{2-}	2000	97.8[a]
PO_4^{3-}	100	104.3[a]
NO_3^-, CO_3^{2-}	400, 400	93.7[a]
SO_3^{2-}	50	56.5[b]
SCN^-	3300	113[a]
S^{2-}	15	58.7[b]
Ca^{2+}, Mg^{2+}, Ba^{2+}	3000, 1000, 50	100[b]
Cu^{2+}	100	52[b]
Ni^{2+}	25	15[b]
Co^{2+}	10	35[b]
Zn^{2+}	100	67[b]
Pb^{2+}	400	5[b]
Ti^{4+}	400	100[b]

[a] Based on a sample containing 0.18 µg of selenium.
[b] Based on a sample containing 0.37 µg of selenium.

Selectivity in immunoassay is understood in terms of cross reactivity or the degree to which the antibody will bind to a substance other than its target analyte. It can be measured by determination of the concentration of another substance that has the same degree of binding as a particular concentration of the target analyte. For example, the radioimmunoassay of barbiturates,[31] using a radioiodinated derivative of 4-hydroxyphenobarbitone, can measure therapeutic levels of barbiturates in very small amounts (50 µl) of blood and urine samples. Selectivity or cross reactivity is measured as the concentration in μg cm^{-3} that has the same degree of binding as 100 ng cm^{-3} of quinalbarbitone (Table 8). From these results, it is apparent that the antibody/antiserum has the necessary characteristics for development of an assay for screening samples for the presence of barbiturates, i.e. a good level of binding of the major members of the series and a very low level of binding of potentially interfering compounds such as primidone or phenytoin. However, primidone is metabolised to phenobarbitone, which will be detected by the assay. The antibody/antiserum also

Table 8 Cross reactivities of barbiturates, barbiturate metabolites and structurally related compounds

Compound	Cross reactivity (μg cm^{-3})	Compound	Cross reactivity (μg cm^{-3})
Quinalbarbitone	0.10	3-Hydroxyamylobarbitone	1.2
Amylobarbitone	0.12	4-Hydroxyphenobarbitone	1.8
Cyclobarbitone	0.14	Glutethimide	> 100
Pentobarbitone	0.16	Phenytoin	> 100
Phenobarbitone	0.26	Primidone	> 100
Barbitone	0.45	Thiobarbituric acid	> 100
N-Methylphenobarbitone	3.8	Caffeine	> 100
3-Hydroxypentobarbitone	0.5	Theophylline	> 100

shows a high level of binding with hydroxylated metabolites in addition to their parent barbiturates. This is a particularly important feature of an assay that is designed for use in forensic toxicology, where urine samples are commonly assayed.

2.3 UNIT PROCESS NO. 3. OBTAINING A REPRESENTATIVE SAMPLE AND ITS MEASUREMENT

2.3.1 Solid Materials

Bulk solid materials such as minerals, foodstuffs, soils and many industrial products contain arbitrary irregular units, not discrete, identifiable constant units. In this situation, the analytical scientist must identify the population from which the sample is to be obtained, select and withdraw valid gross samples of this population and then reduce the gross samples to a laboratory sample(s) suitable for application of the analytical method or technique. If sampling plans are poorly considered, the relationship of the final analytical result to the population from which the sample is withdrawn is uncertain or impossible to interpret. For example, if the material to be assayed is a crushed mineral ore as it enters a smelter via a conveyor belt, then the analytical scientist could obtain a sample representative of a day's run by application of the following method. A scoopful of 500 g of the crushed ore could be removed from the belt every 30 min in a 24-h period to give a total gross sample of 24 kg. This would have to be reduced to a sample of a few grams for assay in such a way that this final sample is thoroughly mixed and uniform and be exactly representative of the larger quantity of material from which it was taken. This is achieved by the method of coning and quartering. First, the gross sample is mixed by shovelling into a succession of cones, removing material from the edges of one cone and shovelling it into the centre of the next cone. The quantity of the sample is then reduced to a manageable size by flattening successive cones and removing alternate quarters to produce cones of smaller sizes until a manageable small portion is obtained. It should be noted that the 500-g samples were amalgamated to form a composite gross sample since it was believed that the 500-g samples were unbiased with respect to the different size and types of particles present in the bulk material. The size of these samples is often a compromise based on the heterogeneity of the bulk material on the one hand and the cost of the sampling operation on the other.

When the relative homogeneity of a material to be sampled is unknown, a good approach is to collect a small number of samples, using experience and intuition as a guide to making them as representative of the population as possible, and then to assay for the component of interest. From these preliminary assays, the standard deviation of the individual samples can be calculated and confidence limits for the average composition can be established using the relationship $\mu = \bar{x} \pm ts/\sqrt{N}$, where μ is the true mean value of the population, $\bar{x}$ is the average of the analytical measurements and t is obtained from statistical tables (Table 13) for N measurements at the desired level of confidence, usually 95%. Table 9 lists some t values from

Table 9 Variation of confidence limits with number of measurements, N

N	t	$ts/\sqrt{N}$ [a]
2	12.71	±18
4	3.18	±3.2
6	2.57	±2.1
8	2.37	±1.7
16	2.13	±1.1
32	2.04	±0.7
100	1.98	±0.4
∞	1.96	0

[a] Based on a 95% confidence that the interval will contain 95% of the samples.

which it can be seen that for $s = 2$ (based on measurement of N samples), approximately six samples are required to obtain a 95% confidence in estimating the true mean.

More refined sampling plans involving, for example, minimum size of individual samples, can then be devised.[32]

An excellent example of the importance of sampling is given in the determination of aflatoxins in peanuts.[33] The aflatoxins are highly toxic compounds produced by moulds that grow best under warm, moist conditions. Such conditions may be localised in a warehouse, resulting in a patchy distribution of highly contaminated kernels. One badly infected peanut (less than 0.1% are contaminated and a single kernel may contain 1000 ppm of aflatoxin) can contaminate a relatively large lot with unacceptable levels (above 25 ppb for human consumption) of aflatoxins after grinding and mixing. The standard deviations of the three operations of sampling, subsampling (which involves grinding/milling of 21.8 kg of sample) and analysis are shown in Figure 8. The analytical method consists of solvent extraction, TLC and measurement of the fluorescence of the aflatoxin spots. Clearly sampling is the major source of the analytical uncertainty.

Sampling of other solid materials can be carried out as follows.[34] Animal feed in a bag can be sampled by insertion of a slotted tube and rod assembly diagonally into the bag lying horizontally and then removing a core the length of the tube. Fresh meat can be sampled by separating it from the bone and passing it three times through a food chopper with 3-mm plate openings, mixing after each chopping step. These steps and the subsequent analysis should be completed as rapidly as possible to avoid loss of moisture and decomposition. Bulk lump lime in railroad cars can be sampled by collection of at least 10 shovelfuls from different parts of each car, which should then be crushed to pass a 5-cm opening. The samples should be mixed and reduced to 1 kg by quartering. This subsample should be stored in a dry, sealed container. All of these steps must be performed as rapidly as possible to avoid the uptake of water.

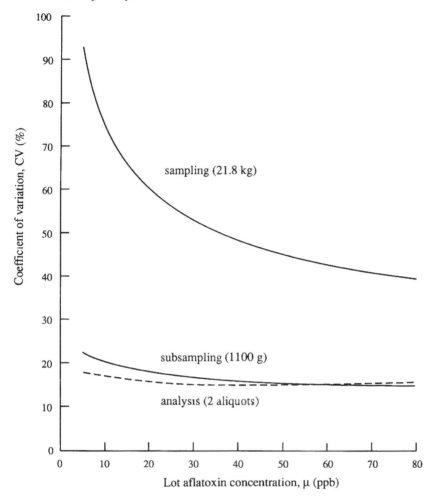

Figure 8 Relationship between lot aflatoxin concentration and coefficient of variation associated with the sampling, subsampling and analytical steps of a peanut aflatoxin testing programme. Reproduced by permission of the author from T. B. Whitaker, *Pure Appl. Chem.*, 1977, **49**, 1709

2.3.2 Liquid Materials

Liquid materials are sampled in a variety of ways followed by volume measurement of a representative sample. Grade A flasks of, say, 1000 cm^3 will measure this volume to ± 0.40 cm^3 ($\pm 0.04\%$); grade A pipettes of 25 cm^3 will measure this volume to ± 0.03 cm^3 ($\pm 0.12\%$); grade B flasks and pipettes will have twice these tolerances. Micrometer syringes will measure volumes in the range 1 cm^3–1 μl with the region of low precision being around 1 cm^3.

Sampling of river water, for example, involves acquisition of a representative sample, noting temperature, time, place, flow rate, etc. In surveying, larger grid patterns are established, with samples being taken at the intersections. Contours, joining sites of identical concentration in a particular pollutant, are constructed as shown in Figure 9 for Zn^{2+}.

Sampling such natural waters for the determination of trace metal ions must be made in containers that are rigorously cleaned (e.g. glass, polyethylene and polypropylene materials must be acid washed and then tested for the metal ion content prior to sampling). Even then contamination during sampling is the most common source of error in environmental trace analysis with the acquisition of inaccurate results of excellent precision. Such contamination can come, for example, from the exhaust of the sampling boat or the line attached to the sampling bottle as it is lowered to the required depth. Most sensitive to contamination errors are usually all types of natural waters, particularly sea waters, in which metal ions such as Pb^{2+} exist at low ppb concentrations compared with ppm levels in some soils and edible vegetables (Table 1).

2.3.3 Gaseous Materials

Gases in an air sample can be sampled by drawing air through an absorbing solution prior to a chemical method of analysis. This method therefore requires that there be a reasonably high concentration of the component to be determined. Interferences are common, e.g. in the determination of SO_2 by reaction with H_2O_2 to yield titratable H_2SO_4, acidic and basic substances in the air sample obviously affect the titre. These methods also involve lengthy sampling periods for sufficient sample to be collected and also result in mean values of concentration over the collection period.

Gas-phase pollutants are generally encountered, together with many potential analytical interferences, at much lower concentrations in workplace atmospheres and the environment. In addition, the pollutant may be found dissolved in aqueous

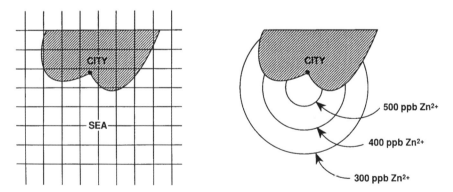

Figure 9 Use of contours to map pollutant Zn^{2+} concentrations in the aqueous environment of a coastal city

aerosols or adsorbed on particulate matter, apart from being present in the gas phase. Sampling pumps are used to draw the sample through a suitable solution for derivatisation (particularly in the case of unstable analytes, e.g. isocyanates, discussed in Section 6.1), or through a combination of filter for particulates and activated carbon or a porous polymer such as Tenax GC for gases. In the latter case, a sampling procedure using, say, activated carbon contained in a sorbent tube can be followed by desorption with a suitable solvent (e.g. CH_3OH, CS_2) and determination of the pollutant(s) by instrumental techniques which possess good selectivity and sufficient LODs for the analyte molecules, e.g. GLC or HPLC. Automated thermal desorption is also used prior to determination by GLC. In some cases it is possible to use a non-separation instrumental 'finish' where the technique is selective enough towards the analyte(s) in the presence of potential interferences from the air sample. The following worked example is illustrative of this, in which spectrofluorimetry is used to determine selectively a polycyclic aromatic hydrocarbon after its collection in a sorbent tube and desorption with pentane.

The carcinogenic hydrocarbon benzo[a]pyrene (V) (MW 252) was concentrated on activated carbon after 50 dm^3 of factory air had been drawn through a sorbent tube containing the adsorbing material for 10 min. The activated carbon was then treated with 100 cm^3 of pentane and the resulting solution was subjected to fluorimetric analysis using the emission at 454 nm. A fluorescence intensity of 11.05 arbitrary units was observed. A series of standard benzo[a]pyrene solutions gave the spectrofluorimetric data in Table 10 when subjected to similar excitation conditions.

Table 10 Fluorescence intensity at 454 nm for standard solutions of benzo[a]pyrene

Benzo[a]pyrene concentration (mol dm^{-3})	Fluorescence intensity (arbitrary units)
1×10^{-8}	1.30
3×10^{-8}	3.90
6×10^{-8}	7.80
1×10^{-7}	13.00

Calculate the concentration of benzo[a]pyrene in the sample of factory air in $\mu g\ m^{-3}$. Assume 100% of the benzo[a]pyrene in the air sample is successively adsorbed and then desorbed prior to spectrofluorimetry.

A plot of fluorescence intensity vs concentration of the standards is linear over the concentration range 10^{-8}–10^{-7} mol dm^{-3}.

(V)

The unknown of fluorescence intensity of 11.05 units corresponds to 8.5×10^{-8} mol dm^{-3}, which is $8.5 \times 10^{-8} \times 252$ g dm^{-3}.
Therefore, in 100 cm^3 of pentane there will be

$$8.5 \times 10^{-8} \times 252 \times 10^{-1} \text{ g} = 2125 \text{ ng or } 2.125 \text{ μg}$$

Hence there are 2.125 μg of benzo[a]pyrene per 50 dm^3 of factory air, which is

$$\frac{2.125}{50} \times 10^3 \text{ μg m}^{-3}, \text{ i.e. } 42.5 \text{ μg m}^{-3}$$

It should be noted that this value is within the OSHA 8-h TWA exposure limit of 200 μg m^{-3}.

In the field of industrial hygiene for the monitoring of workplace atmospheres, a technology of portable and personal sampling devices using sorbent tubes, colour detector tubes, badge methods, etc., has developed. Section 4.6 is concerned with the determination of nitrogen dioxide in air samples by sorbent tube collection–colorimetry in order to assess compliance with the UK COSHH regulations (1994). Further reading material on air pollution analysis which includes a section on gas sampling is available,[35] as is a guide to commercially available devices for air sampling, together with NIOSH/OSHA limits for airborne contaminants and the corresponding analytical methods which relate to workplace exposure.[36]

2.4 UNIT PROCESS NO. 4. PRELIMINARY TREATMENT OF SAMPLE

Although a number of chemical analyses are performed directly on solids (e.g. preparation of a KBr disc prior to application of infrared spectrometry), many measurements are performed on aqueous solutions containing the analyte(s). This therefore means that these solids must be efficiently dissolved prior to application of the analytical technique, e.g. acid digestion of inorganic solids prior to determination by atomic absorption spectrometry. Hot, concentrated HCl works well to dissolve many metal oxides but a number of elements that form volatile chlorides are partially or completely lost from this dissolution medium, e.g. arsenic and antimony trichlorides, tin(IV) and germanium tetrachloride and mercury(II) chloride. Most metal samples with the exception of Al and Cr are dissolved by hot concentrated HNO$_3$. H$_2$SO$_4$ has a very high boiling point of 340°C, a temperature that results in the dissolution of many metals, alloys and biological materials. In the latter case, organic compounds are dehydrated and oxidised by hot concentrated H$_2$SO$_4$, therefore eliminating such components from a sample by what is called sample digestion or wet ashing. In practice, the sample is fumed to charring with H$_2$SO$_4$, followed by dropwise addition of HNO$_3$ to the fuming mass until the charred black material is decolorised. The solution is again heated to fumes and the process is repeated as often as is necessary. A number of elements can be partially or fully volatilised in this procedure, particularly if the sample contains chlorine, e.g. As, B, Ge, Hg, Sb, Se, Sn, S, P and the halogens. Other mixtures of acids can also be used in such wet ashing procedures. Concentrated $HClO_4$ is a very strongly oxidising acid that can be used to dissolve samples as intractable as stainless steel. Oxidising mixtures of acids are often more effective in dissolving oxidisable samples than are the pure acids, e.g. aqua regia [(concentrated HNO_3–concentrated HCl $(1+3)$] can dissolve materials as chemically resistant as Pt metal. HF is used to dissolve silicate minerals with the Si being lost as volatile SiF_4. The remaining HF, which interferes in many chemical analyses, since the fluoride complexes of several cations are extraordinarily stable with different properties from the parent cations, is driven off by evaporation with H_2SO_4 or $HClO_4$.

Some samples, notably those composed of mineral oxides and silicates, are very resistant to aqueous reagents and are best dissolved in molten salt mixtures called fluxes. Fluxes that are used for sample dissolution are normally liquids at 300–1000°C, such high temperatures contributing to their effectiveness in dissolving samples. Fluxes fall into three categories, acidic, basic or oxidising. Potassium pyrosulphate ($K_2S_2O_7$) is a typical acid flux that is used for fusions at ca 400°C and is particularly useful for the attack of the more intractable metal oxides. Na_2CO_3 is a typical basic flux which can be used for silicates and certain other refractory materials by heating to 1000–1200°C; this treatment generally converts the cationic constituents of a sample to acid soluble carbonates or oxides with the non-metallic constituents being converted into soluble Na salts. Such carbonate fusions would normally be carried out in Pt crucibles. Sodium peroxide (Na_2O_2) is a strongly basic oxidising flux that can be used for acid-insoluble alloys of Fe, Ni, Cr, Mo, W and La.

As has already been mentioned, organic matter can be oxidised with an oxidising mineral acid such as H_2SO_4 or $HClO_4$ (digestion/wet ashing). Another means of oxidising an organic sample matrix is by dry ashing in a crucible placed in a muffle furnace. Analysis of the non-volatile components follows solution of the residual solid. Conventional dry ashing carries with it the potential for significant loss of sample through convection currents swirling around the crucible entraining very fine sample particles and carrying them away. In addition, the crucible material is important since many porcelain glazes contain Pb, Zn and other metals, often variable in their concentration. Low-temperature ashing makes use of chemically reactive oxygen species to ash organic samples at slightly above room temperature. The sample is held in an evacuated chamber into which O_2 is bled through a radiofrequency field. Species such as atomic oxygen (O) react with C and organically bound H, S and N and leave the mineral constituents in a largely unaltered form.

The storage, preservation and transport of field samples presents problems of its own. Solid samples can be freeze-dried to reduce microbial activity in the short term or air dried at less than 40°C in the determination of nutrients, K^+, Ca^{2+}, Na^+ and Mg^{2+} and at 105°C for total minerals and total P, S, C and N. Water samples in the field can change quickly in composition owing to chemical and microbial reactions. Cations such as Fe^{2+} and Mn^{2+} can oxidise to their higher oxidation states by air oxidation and precipitate, e.g.

$$4Fe^{2+}(aq) + O_2 + 10H_2O \longrightarrow 4Fe(OH)_3(s) \downarrow + 8H^+$$

Other metal ions whose solubility products are not exceeded can coprecipitate with $Fe(OH)_3$. N, P and S can also undergo relatively rapid chemical change, e.g. NO_3^- is reduced by organic matter/bacteria to give N_2. To minimise these changes, the sample container should be pretreated to leach out ions, the temperature of the sample should be reduced and the addition of preservatives such as $HgCl_2$, addition of H_2SO_4 to pH 1 and addition of solvents such as $CHCl_3$, $C_6H_5CH_3$ and $C_2H_4Cl_2$ are recommended, particularly to slow bacterial action.

After a blood sample has been collected in a clean, dry syringe, it should be transferred to a sample container containing an anticoagulant such as $K_2C_2O_4$ (which binds up the Ca^{2+} as $CaC_2O_4\downarrow$ —the Ca^{2+} initiates the clotting process). This should then be centrifuged to separate the cells (leukocytes, erythrocytes, platelets) from the plasma, which is then subjected to the unit processes of separation and determination. If serum is required, the anticoagulant should be omitted and the blood allowed to clot, prior to separation of the clear serum from cells and fibrinogen by centrifugation. This centrifugation should be carried out as quickly as possible to avoid destruction of the red blood cells by haemolysis and subsequent loss of haemoglobin and other substances such as urea, K^+, Fe^{3+} and Mg^{2+} into the serum, which may interfere with subsequent analyses. The serum sample can then be subjected to protein precipitation with Cl_3CCOOH or $HClO_4$, followed by centrifugation and separation of the supernatant, prior to, say, HPLC.

In drug analysis, plasma and serum are most commonly sampled because a good correlation between drug concentration and therapeutic effect is usually found. Urine analysis for drugs is used in connection with urinary excretion and bioavailability studies. Saliva and cerebrospinal fluid are less commonly analysed. The stability of drugs and their metabolites in biological fluids during storage is of critical importance—if analysis is to be carried out within a few days, storage at 4°C is recommended, and at −20°C for storage for longer periods. Freeze-drying is only recommended for very long-term storage. The sample should be divided into aliquots before freezing to minimise precipitation/degradation by repeated freezing and thawing.

The reactive nature of many organic pollutants and the susceptibility of such species to absorptive and/or adsorptive losses during storage must be considered and where possible minimised. Many gaseous organic pollutants are photochemically unstable so light should be excluded during storage. All possible precautions should be taken against inward or outward leakage of gas samples, particularly during storage. The use of sealing lubricants to prevent leaks is not recommended since these frequently absorb or dissolve gaseous constituents, thus interfering with the sampling process. Condensation of atmospheric moisture within the sampling system should also be avoided for the same reason, and also because moisture may allow certain unwanted reactions to occur.

2.5 UNIT PROCESS NO. 5. SEPARATION OF ANALYTE(S) FROM INTERFERENCES AND EACH OTHER

2.5.1 Separation by Masking

The process of masking involves removal of the chemical reaction or instrumental signal of the interferent prior to determination of the analyte. This is frequently accomplished by introducing a complexing agent that reacts selectively with the interference, e.g. in the iodometric determination of Cu, Fe(III) can be rendered unreactive towards I^- by complexation with F^- or PO_4^{3-}. Neither anion inhibits the oxidation of I^- by Cu(II). Also, CN^- is often employed as a masking agent to permit the titration of Mg^{2+} and Ca^{2+} with EDTA in the presence of ions such as Cd^{2+}, Co^{2+}, Cu^{2+}, Ni^{2+} and Zn^{2+}, all of which form stable cyanide complexes and hence do not complex with EDTA. The interference of Fe(III) in the potentiometric determination of F^- by a fluoride-selective electrode is achieved by complexation of Fe(III) with citrate so that the F^- can be determined in the free form.

2.5.2 Separation by Precipitation

Separation by precipitation is widely used in inorganic quantitative analysis, e.g.

(a) $Cl^- + Ag^+ \rightarrow AgCl\downarrow$
(b) $Ba^{2+} + SO_4^{2-} \rightarrow BaSO_4\downarrow$
(c) $Hg(II), Cu(II), Ag(I) \xrightarrow{S^{2-}} HgS, CuS, Ag_2S\downarrow$

(where S^{2-} is generated by the hydrolysis of thioacetamide

$$CH_3C(=S)NH_2 + H_2O \rightarrow CH_3C(=O)NH_2 + H_2S$$

and these three cations will precipitate in solution of varying acidity from 3 mol dm^{-3} HCl to pH 9).

(d) Twenty-four metal ions form sparingly soluble chelates such as (VI) with 8-hydroxyquinoline (oxine).

Mg^{2+} +

8-Hydroxyquinoline   (VI)

(e)

$$Ni^{2+} + \underset{\text{Dimethylglyoxime}}{H_3C-\underset{\underset{HO}{N}}{\overset{\|}{C}}-\underset{\underset{OH}{N}}{\overset{\|}{C}}-CH_3} \longrightarrow$$

(VII) Ni chelate with dimethylglyoxime — Bright red

Only the Ni chelate (VII) precipitates in a weakly alkaline solution. The bright red precipitate can be dried at 120°C. Ni is quantitatively analysed in steel using such a precipitation procedure (Section 3.10).

Separation by precipitation can also be used in trace analysis by coprecipitation of trace metals with the precipitation of a major metallic constituent in the sample, e.g. the chelating agent 8-hydroxyquinoline together with tannic acid and thioanilide has been used on a soil extract to effect coprecipitation of the major constituent Al together with trace elements Co, Ni, Mo, Sn, Pb, Zn, Cr, V, Ti, Be and Ge. After filtering and ashing, the resulting Al_2O_3 matrix was used for spectrographic determination of the trace metals contained therein with errors $< \pm 10\%$ for quantities of 2–100 μg.[37] This concentration method is useful in ensuring that all spectrographic determinations are made in similar matrices, so eliminating errors due to variations in the major composition of materials such as plant ashes.

2.5.3 Separation by Solvent Extraction

Solvent extraction is widely used not only to separate the analyte(s) from a complex sample which may contain interferences but also to concentrate the analyte(s) prior to the measurement step (unit process No. 6). A basic requirement is that the analyte should be uncharged or form part of an uncharged ionic aggregate. Organic analytes are therefore generally solvent extracted at a pH where they exist as neutral molecules [e.g. carboxylic acids (RCOOH) into diethyl ether] and inorganic analytes solvent extracted after formation of uncharged metal chelates, e.g. copper acetylacetonate (VIII), or overall neutrally charged ion pair complexes, e.g. [Fe(o-phenanthroline)$_3$]$^{2+}$ 2ClO$_4^-$ extractable into chloroform. In both cases, higher recoveries are achieved in solvent extraction by the use of multiple extractions and combining the extracts.

> For example, 10 cm^3 of an aqueous solution containing 1.235 mg of a particular drug was solvent extracted with 5 cm^3 of toluene after which 0.346 mg of the drug remained in the aqueous phase. Given that the distribution coefficient K is given as the concentration of the drug in the organic phase divided by the concentration of the drug

Unit Processes of Analytical Procedures

(VIII)

in the aqueous phase, use equation 11 to calculate (a) how many 10-cm^3 extractions using toluene would be required to recover at least 99% of the drug from 20 cm^3 of the aqueous solution and (b) how many 10-cm^3 would be required to recover at least 99.9% of the drug from 20 cm^3 of the aqueous solution.

$$[\text{drug}]_{aq,n} = \left(\frac{V_{aq}}{V_{org}K + V_{aq}} \right)^n [\text{drug}]_{aq} \qquad (11)$$

where
  $[\text{drug}]_{aq,n}$ is the amount of the drug in the aqueous phase after n extractions
  $[\text{drug}]_{aq}$ is the amount of the drug in the aqueous phase originally
  V_{aq} is the volume of the aqueous phase;
  V_{org} is the volume of the organic phase;
  K is the distribution coefficient; and
  n is the number of solvent extractions.

$$K = \frac{1.235 - 0.346}{5} \bigg/ \frac{0.346}{10}$$
$$= 5.14$$

(a) Using equation 11:

$$1 = \left[\frac{20}{(10 \times 5.14) + 20} \right]^n \times 100$$
$$0.01 = (0.280)^n$$
$$n = \frac{\log 0.01}{\log 0.280} = 3.4$$

Hence four extractions are required.
(b) Again using equation (11):

$$0.1 = \left[\frac{20}{(10 \times 5.14) + 20} \right]^n \times 100$$
$$n = \frac{\log 0.001}{\log 0.280} = 5.4$$

Hence six extractions are required.

To illustrate the solvent extraction of an organic analyte, determination of the diuretic hydrochlorothiazide(IX) in scrum is taken.[38] Serum is first extracted with ethyl acetate to partition the drug into the organic solvent layer. At the pH of serum, **IX** exists as an uncharged neutral molecule since all nitrogen atoms will be unprotonated. The ethyl acetate layer is then back-extracted with 1 mol dm^{-3} NaOH which

$$\text{(IX)}$$

Structure: benzene ring with substituents NH_2SO_2-, Cl-, and a fused 1,2,4-thiadiazine ring containing $S(O_2)$–NH–CH_2–NH.

(IX)

transfers **IX** as an anionic species (a nitrogen atom in the heterocyclic ring can lose a hydrogen ion in alkaline media) back into the aqueous layer. The hydrochlorothiazide is then solvent extracted with diethyl ether and the ether layer is discarded. This triple solvent extraction frees the analyte from certain biological interferences prior to its determination in the remaining aqueous phase by HPLC using a Bondapak C_{18} reversed-phase column with H_2O–MeOH as mobile phase and detection at 271 nm. Compound **IX** elutes at 12 min and an LOD of 50 ng cm^{-3} is quoted. A particular advantage of the method is that seven other drugs, which could be co-administered in a hospital regime together with **IX**, did not interfere.

The determination of trace levels of Cu in a sample of river water is taken as an illustration of solvent extraction of an inorganic analyte.[39] The pH of 100 cm^3 of the water sample is adjusted to within the range 1–6, 5 cm^3 of extractant solution (1 g of ammonium pyrrolidine dithiocarbamate in 100 cm^3 of H_2O) are added and the resulting solution is shaken for 30 s. A 10-cm^3 volume of methyl isobutyl ketone is then added to the solution in a separating funnel and the mixture is shaken vigorously for 2 min. After this solvent extraction, the organic layer, into which an overall neutrally charged Cu dithiocarbamate chelate has been transferred, is withdrawn and aspirated into the flame of an atomic absorption spectrometer. In this example, the concentration factor for the analyte is 10.5, which means that Cu in river water is now detectable by flame atomic absorption spectrometry whereas in the original river water sample it would not have been. Even though many other metals are extracted using these experimental conditions, the excellent selectivity of the Cu hollow-cathode lamp for Cu frees this method from interferences from other metals. Calibration of the method is achieved by extracting all of the Cu from a seawater sample or an artificial sea-water matrix, adding known quantities of Cu to the matrix and extracting these standards as for the sample. A blank should also be determined in the same manner.

2.5.4 Separation by Solid-Phase Extraction

Solvent extraction of organics from biological matrices is, however, subject to the individual operator's technique which can result in varying recoveries and hence inaccurate determinations. In addition, hazardous solvents may be required and solvent extraction is largely unselective towards analytes of similar chemical structure. Intractable emulsions and the need for waste disposal of solvents are further disadvantages. Solid-phase extraction has, in the last decade or so, offered an alternative to such solvent extraction, with fewer disadvantages.[40] Solid-phase extraction uses a solid surface (usually based on a powdered porous silica to which organic functional groups have been bonded) as the extracting phase. Organic molecules in the sample can interact with this solid extracting phase in a variety of ways, e.g. hydrogen bonding and ion pairing. In addition, a porous solid phase can discriminate between molecules of different size. Solid-phase extraction is carried out in disposable cartridges packed with a few hundred milligrams of chemically bonded silica with samples being driven through by positive pressure supplied by a syringe or by the use of a vacuum manifold. There are four stages in solid-phase extraction: conditioning of the sorbent prior to sample application with MeOH, retention of the analyte and other undesired matrix constituents, rinsing to remove some of these interferences and finally elution of desired analyte(s) with a suitable solvent prior to instrumental determination.

The determination of the abused substance cocaine and its metabolites in a urine sample can be taken as an illustration of the use of solid-phase extraction, which is presented in Section 6.5. Such solid-phase extraction procedures can now be automated and Varian have recently introduced the 9200 Prospekt model for this purpose. When linked up on-line for direct injection into the HPLC system, the entire sample from the SPE cartridge is transferred to the head of the column, improving the LOD of the automated method.

2.5.5 Separation by Chromatography

Chromatographic techniques separate multicomponent mixtures based on the different rates at which the components migrate through a stationary phase under the influence of a mobile phase. Not surprisingly, they have been and are being widely used for the analysis of complex matrices with gas–liquid chromatography and high-performance liquid chromatography being particularly popular. The principles of these techniques and other chromatographic techniques such as supercritical fluid chromatography, thin-layer chromatography, ion-exchange chromatography and gel permeation chromatography are discussed in the many technique-orientated texts in analytical chemistry/science, e.g. Refs 41–43. Applications of some of these techniques appear in Chapters 4–6, e.g. the determination of organomercury compounds in fish samples by HPLC–cold vapour AAS (Section 4.2), the determination of organolead compounds in air samples by gas chromatography–AAS (Section 4.5), stability-indicating assay for oxazepam tablets and capsules (Section 5.2) and determination of contaminating antibiotic trace concentrations in dairy feedstuffs by GC–MS (Section 6.4).

2.5.6 Separation by Electrophoresis

Charged species can be separated by electrophoresis owing to differential migration across a surface or through a column in an applied potential gradient. Of particular current importance is capillary electrophoresis, where the separation takes place in a narrow-bore fused-silica capillary tube containing an electrolyte solution. Principles of electrophoresis are discussed elsewhere[41-43] and the application of capillary zone electrophoresis and micellar electrophoretic capillary chromatography to the separation of water soluble vitamins is discussed later (Section 5.3).

2.6 UNIT PROCESS NO. 6. MEASUREMENT

A wide variety of properties of the analytes may be measured, as illustrated in Table 11. Most textbooks dealing with analytical chemistry/science will consider these groups of analytical techniques in depth and the reader is therefore referred to these for detailed information on the measurement step. A few are listed[41-44] with which the author is personally acquainted and has found of value in teaching undergraduate and postgraduate courses, from both theoretical and practical viewpoints.

Effective measurement of complex matrices containing organic and organometallic analytes relies on the use of chromatographic techniques such as HPLC, GLC and, in recent times, CE, allied as coupled techniques with powerful detectors such as mass spectrometers for organics and atomic absorption/emission spectrometers for organometallics. Inorganic species in complex matrices can be determined by a variety of techniques such as atomic absorption spectrometry, inductively coupled plasma spectrometry and stripping voltammetry. Some of these techniques will yield extensive data when applied to a particular complex matrix. Increasing and affordable computational power, which is becoming more intelligent, allows for interpretation of these data, often without the operator needing to have extensive knowledge of it. Since one analytical technique will not necessarily provide all the measurement information to solve the analytical problem totally, there is now an expanding capability to integrate the analytical activities of several instruments

Table 11 Subdivision of analytical techniques and corresponding properties measured. Reproduced by permission of Chapman and Hall from F. W. Fifield and D. Kealey, *Principles and Practice of Analytical Chemistry*, 3rd edn, Blackie, Glasgow, 1990

Group of analytical techniques	Property measured
Gravimetric	Weight of pure analyte or of a stoichiometric compound containing it
Volumetric	Volume of standard reagent solution reacting with the analyte
Spectrometric	Intensity of electromagnetic radiation emitted or adsorbed by the analyte
Electrochemical	Electrical properties of analyte solutions
Radiochemical	Intensity of nuclear radiations emitted by the analyte
Thermogravimetric	Weight changes in a sample as the temperature is varied
Mass spectrometric	Abundance of molecular fragments derived from the analyte
Chromatographic	Physico-chemical properties of individual analytes after separation

through laboratory information management systems (LIMS), automated from sample preparation through to final report. Robotics is assuming an increased presence in these automated procedures, and this is discussed further in Sections 2.10.3, 2.10.4 and 2.10.5.

Apart from the measurement step and, indeed, the whole analytical method being performed by increasingly sophisticated, intelligent and automated instrumentation, there is also the trend of decentralisation of measurement, whereby central laboratories are becoming less important as analysis moves closer to where the qualitative/quantitative result is needed. This is expected to be increasingly the case for the large clinical biochemistry laboratory with the medical doctor making home diagnostic measurements which are accurate and cost effective. It is also expected to be the case using robust, portable, yet accurate and reliable instrumentation for increased on-site environmental monitoring for pesticides, oil pollutants, etc. An example of this latter situation is given in Section 2.10.2 for continuous on-line monitoring of analytes such as Al^{3+} on a river bank by a colorimetric method, with data being sent back to an external computer to be stored and allowing external control of the instrument's programming and operation on the river bank.

2.7 UNIT PROCESS NO. 7. STATISTICAL ASSESSMENT OF MEASUREMENTS

A set of replicate results should number at least 25 if it is to be a truly representative statistical sample. The analytical scientist will rarely consider it economic to make this number of measurements and therefore will need statistical methods to enable him/her to base the assessment on fewer data. First, the reliability of the measurements should be considered and any measurement which deviates substantially from the mean is eliminated by application of statistical tests. This process of data rejection presents the analytical scientist with a paradox. If the limits of acceptance are set too narrowly, results which are rightly part of a statistical sample may be rejected and narrow limits may therefore only be applied with a low confidence of containing all statistically relevant determinations. Conversely, wide limits may be used with high confidence of including all relevant data, but at a risk of including some that have been subject to a gross error. A practical compromise is to set limits at a confidence level of 90% or 95%. There are two criteria which are commonly used to gauge the rejection of results. Of these, the most convenient to use is based on the $\bar{x} + 2\sigma$ interval (where $\bar{x}$ is the mean result and σ the standard deviation) which contains 95.5% of the relevant measurements. Some would believe this limit to be too wide and regard the Q test at the 90% confidence level to be a more acceptable alternative. A rejection coefficient Q is defined as

$$Q = \frac{x_n - x_{n-1}}{x_n - x_1} \tag{12}$$

where x_n is the questionable result in a set $x_1, x_2, x_3, \ldots, x_n$. Q is calculated for the questionable result and compared with a table of critical values (Table 12). The result is rejected if $Q(\text{experimental}) > Q(\text{critical})$.

> For example, the analysis of a dolomite sample yielded $CaCO_3$ percentages of 54.31, 54.36, 54.40, 54.44 and 54.59. The last value appears anomalous—should it be retained or rejected?
> The difference between 54.59 and its nearest neighbour numerically (i.e. 54.44) is 0.15%. The total spread (i.e. $x_n - x_1$) is 0.28%. Therefore,
>
> $$Q(\text{experimental}) = \frac{0.15}{0.28} = 0.54$$
>
> For five measurements $Q(\text{critical})$ at 90% confidence $= 0.64$. Therefore,
>
> $$Q(\text{critical}) > Q(\text{experimental})$$
>
> The result is therefore retained.

Once the reliability of a replicate set of measurements has been established, the mean of the set may be calculated as a measure of the true mean. When this measured value is significantly different from the true mean then the analytical method is inaccurate and is in error, as is illustrated for the determination of Zn in brass below. Absolute error is the difference between the true result and the measured value and relative error is the error expressed as a percentage of the measured value. The true mean will always remain unknown unless an infinite number of measurements is made. A t factor can be used to calculate a confidence interval about the experimental mean, within which there is known confidence (say 90%) of finding the true mean. The limits of this confidence interval are given by

$$\bar{x} \pm ts/\sqrt{N} \tag{13}$$

where $\bar{x}$ is the experimental mean, t is a statistical factor derived from the normal error curve for which tables are available (e.g. Table 13), s is the estimated standard deviation and N is the number of results.

> As an example, the determination of aspirin in an analgesic tablet yields a value of $\bar{x} = 35.40\%$ with a standard deviation of 0.30%. The confidence interval, which varies

Table 12 Critical values of Q at the 90% confidence level

No. of results	$Q(\text{critical})$; 90% confidence
2	—
3	0.94
4	0.76
5	0.64
6	0.56
7	0.51
8	0.47
9	0.44
10	0.41

Unit Processes of Analytical Procedures

Table 13 Table of t-distribution

Degrees of freedom	Confidence (probability) level (%)						
	80	90	95	98	99	99.5	99.9
1	3.08	6.31	12.71	31.82	63.66	127.32	636.62
2	1.89	2.92	4.30	6.97	9.92	14.09	31.60
3	1.64	2.35	3.18	4.54	5.84	7.45	12.92
4	1.53	2.13	2.78	3.75	4.60	5.60	8.61
5	1.48	2.02	2.57	3.37	4.03	4.77	6.87
6	1.44	1.94	2.45	3.14	3.71	4.32	5.96
7	1.42	1.89	2.37	3.00	3.50	4.03	5.41
8	1.40	1.86	2.31	2.90	3.36	3.83	5.04
9	1.38	1.83	2.26	2.82	3.25	3.69	4.78
10	1.37	1.81	2.23	2.76	3.17	3.58	4.59
11	1.36	1.80	2.20	2.72	3.11	3.50	4.44
12	1.36	1.78	2.18	2.68	3.06	3.43	4.32
13	1.35	1.77	2.16	2.65	3.01	3.37	4.22
14	1.35	1.76	2.14	2.62	2.98	3.33	4.14
15	1.34	1.75	2.13	2.60	2.95	3.29	4.07
∞	1.28	1.65	1.96	2.33	2.58	2.81	3.29

inversely with the number of measurements made, for five measurements is $35.40 \pm 2.13(0.30)/\sqrt{5}$, i.e. $35.40 \pm 0.29\%$. Five determinations at most are therefore required to obtain a reasonable estimate of the true mean.

By comparing an experimental mean ($\bar{x}$) with the true value (μ) and using this confidence interval, it is possible to discriminate between determinate errors (i.e. bias in the analytical method due to a regular methodological error) and indeterminate errors.

For example, a new atomic absorption method for the simultaneous determination of Pb and Zn is tested against a brass specimen with an established composition of Pb (0.106% w/w) and Zn (32.20% w/w). The new method gives 0.103% w/w Pb with standard deviation 0.012% and 32.48% w/w Zn with standard deviation 0.10%. If these values represent the average of four determinations, is the existence of determinate error demonstrated in the determination of Pb at the 99% confidence level or in the determination of Zn at the 90% confidence level?

For Pb, the confidence interval $= (5.84 \times 0.012)/\sqrt{4} = 0.035$.

since
$$\mu - \bar{x} = 0.106 - 0.103 = 0.003$$

then
$$|\mu - x| < \left|\frac{ts}{N^{1/2}}\right|$$

i.e. no determinate error is demonstrated.

For Zn, the confidence interval $= (2.35 \times 0.10)/\sqrt{4} = 0.12$.

since
$$\mu - \bar{x} = 32.20 - 32.48 = -0.28$$

then
$$|\mu - \bar{x}| > \left|\frac{ts}{N^{1/2}}\right|$$

i.e. determinate error is demonstrated 90 times out of 100.

When a comparison of two separate replicate sets of data is required, the first stage is normally to compare their respective precisions by means of the F test. F is calculated from $F = S_x^2/S_y^2$ where, by convention, $S_x^2 > S_y^2$ and where S_x^2 and S_y^2 are the respective variances of methods x and y. The F value is compared with critical values (Table 14) calculated on the assumption that they will be exceeded on a probability basis in only 5% of cases using a confidence level of 95%. When F(experimental) > F(critical), then the difference in variance or precision is statistically significant.

> For example, the skills of two analysts are compared by having each perform identical replicate nitrogen analyses on a pure organic drug compound. The relative standard deviation of six analyses obtained by one analyst was 0.13% N and the relative standard deviation of five analyses obtained by the second analyst was 0.07% N. Does this difference in precision suggest a difference in abilities between the two analysts?

$$F(\text{experimental}) = \frac{0.13^2}{0.07^2} = 3.44$$

F(critical) for five degrees of freedom in the numerator and four degrees of freedom in the denominator is 6.26 using a confidence level of 95%. Therefore, since F(experimental) < F(critical), the skills of the two analysts, with respect to their precisions, are comparable.

The second stage is to compare the mean results from the two sets, using the t test in one of its forms. If there is an accepted value for the result based on extensive previous analysis, t(experimental) is calculated from

$$t = [(\bar{x} - \mu)/s]N^{1/2} \tag{14}$$

where $\bar{x}$ is the mean result of the set, μ is the accepted value, s is the experimental standard deviation and N is the number of results. If there is no accepted value and two experimental means are to be compared, t(experimental) is calculated from (with $M + N - 2$ degrees of freedom)

$$t = [(\bar{x} - \bar{y})/s][MN/(M+N)]^{1/2} \tag{15}$$

where $\bar{x}$ is the mean of M determinations, $\bar{y}$ is the mean of N determinations and s is the pooled standard deviation. This latter equation is a simplified expression derived on the assumption that the precisions of the two sets of data are not significantly different. Thus the application of the F test is a prerequisite for its use. In either case, if t(experimental) > t(critical) for the appropriate number of degrees of freedom, the difference between the means is said to be significant.

To illustrate the application of such Q, F and t tests to a single example, the following is chosen.

> The accepted value for the fluoride content of a standard sample is 54.20% with a standard deviation of 0.15%. Five analyses of the same sample by an alternative analytical method yield results of 54.01, 54.20, 54.05, 54.27 and 54.13%. The question is whether the new method is giving results consistent with the accepted value. Application of the Q test to the five results shows no unreliable results. The mean result

is calculated as $\bar{x} = 54.13\%$ with standard deviation $s = 0.106\%$. Calculation of the F value gives $F(\text{experimental}) = 2.01$. $F(\text{critical})$ for an infinite number of degrees of freedom in the numerator and four in the denominator is 5.63 for a confidence level of 95%. Hence $F(\text{experimental}) < F(\text{critical})$, i.e. the standard deviations are not significantly different. The means are then compared using equation 14:

$$t = [(54.20 - 54.13)/0.106]5^{1/2}$$
$$= 1.47$$

For 90% confidence, $t(\text{critical})$ is found to be 2.13 and hence the alternative analytical method is giving a mean value which is not statistically significantly different from the accepted mean value.

2.8 UNIT PROCESS NO. 8. CALCULATION OF ANALYTICAL RESULT AND SOLUTION TO PROBLEM

2.8.1 Quantitative Analysis

From the viewpoint of quantitative analysis, this unit process refers to conversion of the mean measurement obtained from unit process No. 7 to concentration of the analyte by use of a conventional calibration curve, standard addition method or internal standard method. The internal normalisation method (Section 2.8.1.4) is used to establish the proportions of two or more analytes in a sample.

2.8.1.1 CALIBRATION CURVES

A calibration curve is a plot of the response of the instrument (such as absorbance for electromagnetic radiation absorption, electrochemical cell current or chromatographic peak area) versus concentration, mass or volume of the analyte. Where the sample matrix or reagents contribute significantly to the response, this blank is subtracted from the total response to give the response of the analyte alone. Data are usually plotted directly as, say, current versus concentration, but in some cases logarithmic functions or ratios are also used. A linear relationship between the variables is normally sought, but curved calibration plots may be acceptable if understood and reproducible. Some modern instruments employ automatic curvature correction to extend the linear calibration range. When the linear relationship is established and the scatter of points is not appreciable, then a 'best fit' may be obtained by eye. If the scatter of points is appreciable owing to indeterminate errors associated with the measurement process, then linear regression analysis should be used to obtain a line of 'best fit.'

Linear regression analysis involves computing a slope and an intercept that define this regression line, the line representing the average relationship between the two plotted variables. In addition to providing the working calibration curve, the degree of scatter of the plotted points around the regression line indicates the precision of the calibration data and is normally quoted as a standard deviation about the line.

Table 14 Table of F-distribution

Confidence (probability) level, %	Denominator	Degrees of freedom — Numerator											
		1	2	3	4	5	6	7	8	9	10	15	∞
90	1	39.9	49.5	53.6	55.8	57.2	58.2	58.9	59.4	59.9	60.2	61.2	63.3
95		161.4	199.5	215.7	224.6	230.2	234.0	236.8	238.9	240.5	241.9	246.0	254.3
99		4,052	4,999	5,403	5,625	5,764	5,859	5,928	5,981	6,023	6,056	6,157	6,366
90	2	8.53	9.00	9.16	9.24	9.29	9.33	9.35	9.37	9.38	9.39	9.42	9.49
95		18.5	19.0	19.2	19.2	19.3	19.3	19.4	19.4	19.4	19.4	19.4	19.5
99		98.5	99.0	99.2	99.3	99.3	99.3	99.4	99.4	99.4	99.4	99.4	99.5
90	3	5.54	5.46	5.39	5.34	5.31	5.28	5.27	5.25	5.24	5.23	5.20	5.13
95		10.1	9.55	9.28	9.12	9.01	8.94	8.89	8.85	8.81	8.79	8.70	8.53
99		34.1	30.8	29.5	28.7	28.2	27.9	27.7	27.5	27.3	27.2	26.9	26.1
90	4	4.54	4.32	4.19	4.11	4.05	4.01	3.98	3.95	3.94	3.92	3.87	3.76
95		7.71	6.94	6.59	6.39	6.26	6.16	6.09	6.04	6.00	5.96	5.86	5.63
99		21.2	18.0	16.7	16.0	15.5	15.2	15.0	14.8	14.7	14.5	14.2	13.5
90	5	4.06	3.78	3.62	3.52	3.45	3.40	3.37	3.34	3.32	3.30	3.24	3.10
95		6.61	5.79	5.41	5.19	5.05	4.95	4.88	4.82	4.77	4.74	4.62	4.36
99		16.3	13.3	12.1	11.4	11.0	10.7	10.5	10.3	10.2	10.1	9.72	9.02

6	90	3.78	3.46	3.29	3.18	3.11	3.05	3.01	2.98	2.96	2.94	2.87	2.72
	95	5.99	5.14	4.76	4.53	4.39	4.28	4.21	4.15	4.10	4.06	3.94	3.67
	99	13.7	10.9	9.78	9.15	8.75	8.47	8.26	8.10	7.98	7.87	7.56	6.88
7	90	3.59	3.26	3.07	2.96	2.88	2.83	2.78	2.75	2.72	2.70	2.63	2.47
	95	5.59	4.74	4.35	4.12	3.97	3.87	3.79	3.73	3.68	3.64	3.51	3.23
	99	12.2	9.55	8.45	7.85	7.46	7.19	6.99	6.84	6.72	6.62	6.31	5.65
8	90	3.46	3.11	2.92	2.81	2.73	2.67	2.62	2.59	2.56	2.54	2.46	2.29
	95	5.32	4.46	4.07	3.84	3.69	3.58	3.50	3.44	3.39	3.35	3.22	2.93
	99	11.3	8.65	7.59	7.01	6.63	6.37	6.18	6.03	5.91	5.81	5.52	4.86
9	90	3.36	3.01	2.81	2.69	2.61	2.55	2.51	2.47	2.44	2.42	2.34	2.16
	95	5.12	4.26	3.86	3.63	3.48	3.37	3.29	3.23	3.18	3.14	3.01	2.71
	99	10.6	8.02	6.99	6.42	6.06	5.80	5.61	5.47	5.35	5.26	4.96	4.31
10	90	3.29	2.92	2.73	2.61	2.52	2.46	2.41	2.38	2.35	2.32	2.24	2.06
	95	4.96	4.10	3.71	3.48	3.33	3.22	3.14	3.07	3.02	2.98	2.85	2.54
	99	10.0	7.56	6.55	5.99	5.64	5.39	5.20	5.06	4.94	4.85	4.56	3.91
15	90	3.07	2.70	2.49	2.36	2.27	2.21	2.16	2.12	2.09	2.06	1.97	1.76
	95	4.54	3.68	3.29	3.06	2.90	2.79	2.71	2.64	2.59	2.54	2.40	2.07
	99	8.68	6.36	5.42	4.89	4.56	4.32	4.14	4.00	3.89	3.80	3.52	2.87
∞	90	2.71	2.30	2.08	1.94	1.85	1.77	1.72	1.67	1.63	1.60	1.49	1.00
	95	3.84	3.00	2.60	2.37	2.21	2.10	2.01	1.94	1.88	1.83	1.67	1.00
	99	6.63	4.61	3.78	3.32	3.02	2.80	2.64	2.51	2.41	2.32	2.04	1.00

The regression line can be defined using the method of least squares and a simple computer program will enable the computations to be made rapidly. Adherence to linearity is achieved through the nearness of the correlation coefficient, r, to unity. For $r > 0.99$, linearity is regarded as excellent, as is the case for capillary electrophoresis of the boron–Azomethine H anionic chelate as a technique for the determination of boron (Section 4.4).

It is common practice in the analytical literature to compare two methods by plotting the quantitative results of one method against those of the other. A correlation coefficient, r, of near unity is indicative of good agreement between the methods, but it is also necessary to have a slope of near unity and a minimal intercept on the y axis. If there was a regular determinate error in one of the methods, then r would be near unity but the slope would be unlikely to be so.

Calibration curves are widely used for quantitative analysis. For example, in this text, emission intensity corrected for the background signal is plotted against concentration of Na standards for the determination of Na in mineral waters (Section 3.1); the visible absorbance of the peroxide complex of Ti is plotted against concentration of Ti standards for the determination of Ti in rocks (Section 3.4); the absorbance of the Fe^{3+}–o-salicylic acid chelate at 510 nm is plotted against the concentration of the sodium salt of o-salicylic acid standards for the determination of aspirin in an analgesic formulation (Section 5.1.2). If a linear relationship between instrumental response and concentration has already been proved over a particular concentration range, then only two measurements need be made, one on the sample and one on the standard, for quantitative analysis, i.e.

$$\frac{\text{response of analyte in sample}}{\text{concentration of analyte in sample}} = \frac{\text{response of analyte in standard}}{\text{concentration of analyte in standard}} \quad (16)$$

This is exemplified by a worked example in Section 6.1.4 for the determination of an isocyanate in a factory air sample following its derivatisation with 1-(2-methoxyphenyl)piperazine and HPLC with electrochemical detection.

2.8.1.2 Standard Addition Method

The standard addition method of calibration is used to prepare a calibration plot in cases where the composition of the sample matrix is variable or unknown so that a reagent/sample matrix blank response cannot be reliably subtracted from each standard to arrive at the analyte response alone as with calibration plots. In these cases, the sample is spiked with increasing amounts of analyte. The responses of unspiked and spiked samples are measured and a calibration curve is plotted as shown in Figure 10.

The extrapolation in Figure 10 will give the mass, volume or concentration of the analyte, provided that no significant dilutions of the samples occurred on spiking or that the unspiked and spiked samples were diluted by the same amount. If this is not the case, appropriate mathematical corrections must be applied to the responses.

Unit Processes of Analytical Procedures 59

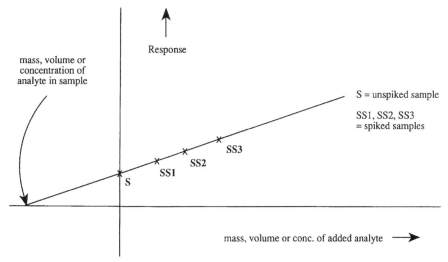

Figure 10 Graphical representation of standard addition method of quantitation. Reproduced by permission of Chapman & Hall from D. Kealey, *Experiments in Modern Analytical Chemistry*, Blackie, Glasgow, 1986

Worked examples of the standard addition method are given later, e.g. the determination of Cu in sea water by anodic stripping voltammetry (Section 4.1.4) and the visible spectrometric determination of boron in plants (Section 4.4.4).

2.8.1.3 INTERNAL STANDARD METHOD

The internal standard method of calibration involves the addition of a constant mass, volume or concentration of a selected standard (which is not the analyte itself) to all samples and to one or more analyte standards. For each sample and analyte standard, the responses of analyte and added internal standard are measured and the ratios of these responses calculated. If several analyte standards are treated in this way, a calibration curve can be plotted of response ratio versus mass, volume or concentration of analyte. It should be noted that the response ratios are independent of the absolute values of the responses themselves, i.e. the sample size. The compositions of unknown samples can then be read from the calibration curve using their calculated response ratios. If a single analyte standard is prepared, then

$$\frac{\text{mass, volume or concentration of analyte in sample}}{\text{mass, volume or concentration of analyte in standard}} = \frac{\text{response ratio for sample}}{\text{response ratio of analyte standard}} \quad (17)$$

Such response ratios are ideally independent of sample size and are usually much less sensitive to variations in experimental conditions than the response of the analyte itself.

As an example of the application of the internal standard method, the reader is referred to the analytical problem in Section 6.7, i.e., the determination of Prozac (fluoxetine) and its demethylated metabolite (norfluoxetine). Following addition of a constant volume of a standard solution of the internal standard, protriptyline, to 0.5-cm^3 aliquots of all serum samples prior to solid-phase extraction, application of reversed-phase HPLC with UV detection gave fluoxetine/protriptyline response ratios for the patient's sample and serum standard of 0.52 and 0.50, respectively. The norfluoxetine/protriptyline response ratios for the patient's sample and serum standard were 0.208 and 0.65, respectively. The reader should refer to Figure 52B (serum standard) and Figure 52C (patient's sample) for visualisation of these data.

Applying equation 17:

$$\frac{\text{concentration of fluoxetine in sample}}{\text{concentration of fluoxetine in standard}} = \frac{\text{response ratio for sample}}{\text{response ratio for standard}}$$

$$\text{Concentration of fluoxetine in sample} = 250\left(\frac{0.52}{0.50}\right) \text{ ng cm}^{-3}$$

$$= 260 \text{ ng cm}^{-3}$$

$$\frac{\text{concentration of norfluoxetine in sample}}{\text{concentration of norfluoxetine in standard}} = \frac{\text{response ratio for sample}}{\text{response ratio for standard}}$$

$$\text{Concentration of norfluoxetine in sample} = 250\left(\frac{0.208}{0.65}\right) \text{ ng cm}^{-3}$$

$$= 80 \text{ ng cm}^{-3}$$

2.8.1.4 INTERNAL NORMALIZATION METHOD

The internal normalisation method of calibration is used when only the proportions of two or more analytes in a sample are to be determined. The composition of the sample is calculated by expressing the response for each component, corrected for differences in sensitivity if necessary, as a percentage of the summed responses for all components. Thus the internally normalised composition of a three-component mixture separated by GLC or HPLC would be established by measuring the peak area for each and expressing it as a percentage of the summed areas, i.e.,

$$\text{component A(\%)} = \frac{\text{peak area for A}}{\text{sum of areas of A, B and C}} \times 100 \qquad (18)$$

For example, the relative peak areas obtained from a gas chromatogram of a mixture of ethyl acetate, ethyl propionate and ethyl n-butyrate were 17.1, 42.6 and 28.9, respectively. Calculate the percentage of each compound in the mixture if the respective relative detection responses were 0.60, 0.78 and 0.88.

The corrected area for ethyl acetate is $17.1 \times 0.60 = 10.26$
The corrected area for ethyl propionate is $42.6 \times 0.78 = 33.23$
The corrected area for ethyl n-butyrate is $28.9 \times 0.88 = 25.43$

Unit Processes of Analytical Procedures 61

The total corrected area = 68.92

$$\text{Ethyl acetate (\%)} = \frac{10.26}{68.92} \times 100 = 14.89$$

$$\text{Ethyl propionate (\%)} = \frac{33.23}{68.92} \times 100 = 48.22$$

$$\text{Ethyl n-butyrate (\%)} = \frac{25.43}{68.92} \times 100 = 36.89$$

2.8.2 Qualitative Analysis

For identification of an analyte, a variety of analytical parameters can be used, some of which are listed in Table 15. The retention time in gas chromatography–electron capture detection as an aid to organochlorine identification of the pesticide dieldrin in a suspect roller meal sample[45] is given in Figure 11. The use of diode-array detection coupled with capillary electrophoresis to confirm more fully the identification of a migrating molecule is illustrated in Figure 12 for the identification of four separated 1,4-benzodiazepines using the MECC technique.[46] The UV spectra of nitrazepam, chlordiazepoxide, diazepam and flurazepam recorded on-line match those recorded for the four molecules using conventional UV spectrophotometry. Figure 13 shows the computer-generated electron impact mass spectrum of benzamide (**X**) as a powerful means of its identification. The fragmentation pattern is given in Figure 14.[47]

2.9 ROLE OF COMPUTERS AND MICROPROCESSORS IN MODERN ANALYTICAL METHODS

It is worthwhile at this stage to mention the important role of computers and microprocessors in modern analytical methods.

2.9.1 Instrument Operation

Computers and microprocessors can help in the initial selection of instrumental parameters entered by keyboard. For example, adsorptive stripping voltammetry, using the Metrohm 646 VA Processor and 647 VA Stand, of the mixture of six textile dyes discussed in Section 2.2 yields the voltammogram shown in Figure 15 using 'keyed-in' instrumental parameters such as deaeration/stirring for 300 s, selection of the hanging mercury drop electrode (HMDE) and the adsorption of the analytes at this electrode for 300 s, stripping from 0 to -1.050 V using the differential-pulse mode of operation with 50-mV pulses applied for intervals of 200 ms. It should be noted that from the viewpoint of selectivity, a separation such as that shown in Figure 15 is inferior to that displayed by HPLC and CE in Section 2.2 since Remazol

Table 15 Some analytical parameters used in the identification of analytes

Parameter	Analytical technique	Comments
Absorption maximum, λ_{max}	UV–visible spectrophotometry	Of limited value in identification of organic molecules and metal chelates
Peak potential	Differential-pulse polarography	Of some value in the identification of electroactive metal ions and organic molecules
Wavelengths and intensities of infrared absorption bands	Infrared spectrometry	Very powerful tool for the identification of organic molecules
Wavelength of absorption	Atomic absorption spectrometry	Of particular value in the identification of many elements in the Periodic Table
Retention time	Chromatography	Of particular value in the identification of 'suitable' organic molecules, e.g. those that are volatile in gas chromatography following separation from like molecules and unwanted matrix components. When retention time can be coupled with spectroscopic identification of the eluting molecule, e.g. GC–IR, GC–MS, HPLC–diode-array UV–visible detection, HPLC–MS, then identification is more assured
Migration time	Capillary electrophoresis	Of particular value in the identification of charged and uncharged organic molecules and metal species following separation from like molecules/species and unwanted matrix components. Coupling with diode-array UV–visible detection and MS is used to assure identification
Fragmentation pattern	Mass spectrometry	Very powerful identification tool for pure samples of organic molecules

Blue GG, Remazol Yellow RNL and Remazol Black B all contribute to the adsorptive stripping signal at -246 mV, Remazol Red RB and Cibacron Red C-2G to the signal at -341 mV and Cibacron Orange CG is responsible for the signal at -476 mV.[48]

The storage and recall of particular groups of instrumental settings for the determination of particular analytes in particular matrices is an advantage of this and

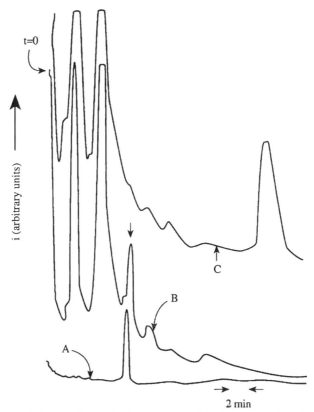

Figure 11 GLC–electron-capture detection of standard dieldrin and selected meal extracts. (A) Standard 2 ppm dieldrin; (B) suspect roller meal sample; (C) breakfast meal sample; 30-g meal samples were extracted with hexane for 48 h using Soxhlet extraction. The extracts were evaporated to less than 10 cm^3 and the volumes made up to 10 cm^3 with hexane prior to GLC-ECD. Reproduced with permission of Springer Verlag from *Fresenius' J. Anal. Chem.*, 1993, **345**, 704. Copyright 1993 Springer-Verlag

other microprocessor-controlled analytical instrumentation such as modern spectrometers and chromatographs. In many cases, a number of analytes can be determined in this way in sequence according to a predetermined programme. This is illustrated (Figure 16) in the sequential determination of trace concentrations of seven metals in electronic-grade material using the aforementioned microprocessor-controlled stripping voltammetry. The predetermined programme is given in Figure 17. The overall method can be automated with a turntable (sample changer) and automatic pipettes (Dosimats).[49]

Furthermore, greater stability of operation can result from the continuous monitoring of instrumental settings. The condition of the various components of an instrument can be continuously monitored by in-built self-diagnostic software so that any deterioration in performance can be brought to the attention of the operator.

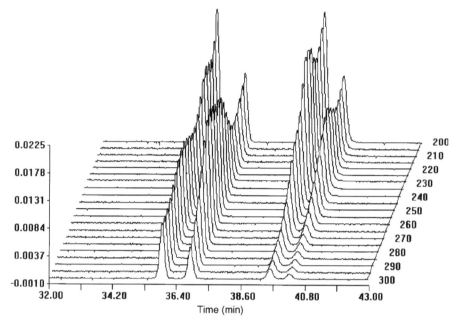

Figure 12 MECC separation of four benzodiazepines at pH 8 using 75 mmol dm^{-3} sodium dodecyl sulphate in 6 mmol dm^{-3} sodium tetraborate–12 mmol dm^{-3} sodium dihydrogenphosphate with 5% methanol. In order of migration: nitrazepam, chlordiazepoxide, diazepam and flurazepam. Reproduced with permission of Elsevier Science from McGrath et al., J. Chromatogr., in press

2.9.2 Data Recording and Storage

The introduction of computers and microprocessors has meant that measurements can now be made significantly faster, e.g. fast Fourier transform IR spectra can be displayed on a VDU screen in seconds and sequential spectrochemical analyses can be performed for ten or more elements in a matter of minutes. These rapidly recorded measurements can then be stored and displayed when necessary in seconds.

2.9.3 Data Processing and Analysis

'Raw' analytical measurements are usually refined by the use of background correction, spectrum smoothing, removing one component from a spectrum, peak area measurement in chromatography, etc., prior to qualitative or quantitative decisions being made by the computer or microprocessor. Mass, NMR and IR spectrometry, in the identification of an 'unknown' molecule, use computer libraries with large numbers of reference spectra to carry out a search for the most likely identity of the unknown and will present the analytical scientist with a limited number of possibilities. The computer or microprocessor in these cases does not replace the skill and experience of the analytical scientist but carries out routine preliminaries enabling

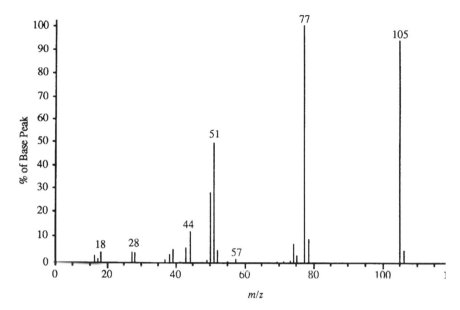

Figure 13 Computer-generated electron impact mass spectrum of benzamide (**X**) in bar graph form. Peak abundances as % of base peak (100%) are reported versus mass to charge (m/z). Peaks of $< 0.5\%$ of the base peak are omitted. A Hewlett-Packard HP 5995 96A GC–MS instrument was used. Reproduced with permission of John Wiley & Sons, Inc., from Silverstein, Bassler and Morrill, *Spectrometric Identification of Organic Compounds*, 5th edn, Wiley, New York

him/her to concentrate his/her efforts on the final interpretation. In almost all cases, quantitative analysis is carried out by calibration with standards. A computer or microprocessor can store these calibration data and hence automatically evaluate routine quantitative measurements immediately.

2.9.4 Validation Testing

Analytical method validation and system suitability testing, as illustrated in Figure 2, if performed and documented manually on a day-to-day basis, consume large amounts of time and operator involvement. Software, such as the Hewlett-Packard HPLC and CE ChemStation, allows for this to be carried out rapidly with the operator only needing to put vials of a selection of standards in the autosampler tray and to start the test procedure. The system injects the standards, collates data from several runs and calculates and prints statistics on precision (of peak areas, peak heights or amounts and retention/migration times), linearity, resolution between p-eaks, etc., in a format widely accepted by regulatory authorities and independent auditors. Table 16 shows the criteria available in the software and which can be determined for a single run or for multiple runs.

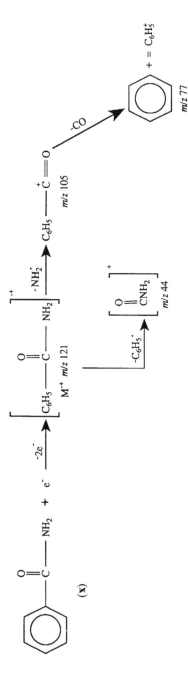

Figure 14 Mass spectral fragmentation pattern of benzamide (**X**). Reproduced with permission of John Wiley & Sons, Inc., from Silverstein, Bassler and Morrill, *Spectrometric Identification of Organic Compounds*, 5th edn, Wiley, New York

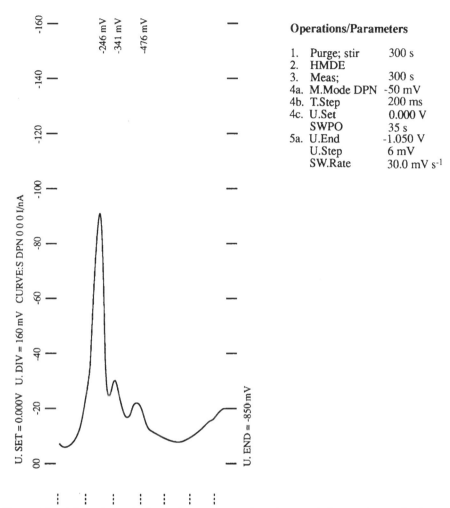

Figure 15 Adsorptive stripping voltammogram of a mixture of six reactive textile dyes. Reproduced by permission of the Royal Society of Chemistry from D. A. Oxspring *et al.*, *Analyst*, 1995, **120**, 1995

2.9.5 Software for Method Development

The comments in Section 2.9.1 refer to the use of computers and microprocessors in selecting already optimised instrumental parameters. Software is now available to assist the analytical scientist in optimising these parameters. For example, in reversed-phase high-performance liquid chromatography, Crome Dream-2 (Knauer) is a program that can support selection of the initial conditions after information on chemical structures has been entered. Capacity factors (k') are calculated from their mathematical relationship with physico-chemical data such as solvation energy of

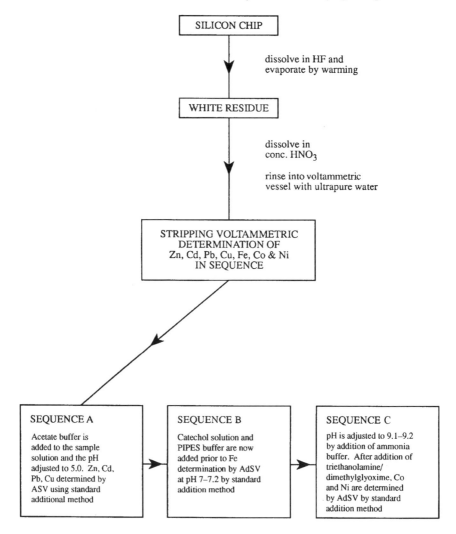

Figure 16 Analytical method used in the sequential determination of seven metals in electronic-grade material using microprocessor-controlled stripping voltammetry

the solutes with water and plots are displayed of ln k' against the concentration of the organic solvent in the mobile phase for each analyte in the sample. An acceptable value of this latter parameter can then be selected when the minimum and maximum retention times are entered. A simulated chromatogram of the mixture is then displayed. Subsequent steps involve the use of real retention data to refine the retention model and to optimise the resolution of the mixture. This can be applied to both isocratic elution and linear and multi-segment gradient elution.

Unit Processes of Analytical Procedures 69

```
Zn,Cd,Pb,Cu,Fe,Ni und Co im sub-ppb-Bereich      METHOD  4  PAGE 3
MPL 1            EL.TYPE MME                     OPERATION SEQUENCE

    OPERATIONS/PARAMETERS                        OPERATIONS/PARAMETERS
 1  PURGE ;STIR  ;CADDL ;  300   s          8a   U.END              -210  mV
 2  PURGE ;STIR  ;         30   s           8b   U.STEP                4  mV
 3  OPURGE;<REP  ;         10   s                SW.RATE             6.6 mV/  s
 4  HMDE  ;STIR  ;MEAS  ;  120   s          9    SWP 3 ;              61  s
 4a      M.MODE        DPN   50   mV        9a   U.END               200  mV
 4b      T.STEP              600  ms        9b   U.STEP                4  mV
 4c      U.SET              -1.109 V             SW.RATE             6.6 mV/  s
 5  OSTIR ;                 10   s         10    REP > 1;BEEP  ;ADD1J2;
 6  SWP 0 ;                 53   s         11    BEEP  ;PURGE ;STIR  ;  30  s
 6a      U.END              -750  mV       12    HOLD  ;CADDL ;
 6b      U.STEP              4    mV       13    PURGE ;STIR  ;         30  s
         SW.RATE             6.6 mV/  s    14    OPURGE;<REP  ;         10  s
 7  SWP 1 ;                 45   s         15    HMDE  ;STIR  ;MEAS ;  120  s
 7a      U.END              -450  mV       15a      M.MODE        DPN  -50  mV
 7b      U.STEP              4    mV       15b      T.STEP              600 ms
         SW.RATE             6.6 mV/  s    15c      U.SET              -250 mV
 8  SWP 2 ;                 36   s         16    OSTIR ;                15  s


Zn,Cd,Pb,Cu,Fe,Ni und Co im sub-ppb-Bereich      METHOD  4  PAGE 3
MPL 1            EL.TYPE MME                     OPERATION SEQUENCE

    OPERATIONS/PARAMETERS                        OPERATIONS/PARAMETERS
17  SWP 4 ;                 60   s         25    SWP 5 ;                60  s
17a      U.END              -650  mV       25a      U.END            -1.000 V
17b      U.STEP              4    mV       25b      U.STEP              2   mV
         SW.RATE             6.6 mV/  s             SW.RATE            3.3 mV/  s
18  REP > 1;BEEP  ;ADD2J2;                 26    SWP 6 ;                37  s
19  BEEP  ;PURGE ;STIR  ;   30   s         26a      U.END            -1.250 V
20  HOLD  ;CADDL ;                         26b      U.STEP              4   mV
21  FURGE ;STIR  ;          30   s                  SW.RATE            6.6 mV/  s
22  OPURGE;<REP  ;                         27    REP > 1;BEEP  ;ADD3J2;
23  HMDE  ;STIR  ;MEAS  ;   30   s         28    OMEAS ;BEEP  ;END    ;
23a      M.MODE        DPN  -75   mV
23b      T.STEP              600  ms
23c      U.SET              -700  mV
24  OSTIR ;MEAS  ;          10   s
24a      M.MODE        DPN  -75   mV
24b      T.STEP              600  ms
24c      U.SET              -800  mV
```

Figure 17 Operation sequence in the determination of seven metals in electronic-grade material using microprocessor-controlled stripping voltammetry. Reproduced by permission of Metrohm AG from *Metrohm Application Note No. 147e*, Metrohm, Herisau, 1987

Table 16 Selected parameters for automated method validation and system suitability testing in the HPLC^{2D} and HPLC^{3D} ChemStation and ^{3D}CE ChemStation software. Reproduced by permission of Hewlett-Packard GmbH, Waldbronn, Germany, from *Good Laboratory Practice—Part 2: Automating Method Validation and System Suitability Testing*, Product Note, 12-5091-6082E

Parameter	HPLC	CE
For a single run		
Plate number of a column	✓	✓
Capacity factor	✓	Not applicable
Peak width at half-height	✓	Not applicable
Tailing factor	✓	Not applicable
Resolution between two peaks	✓	✓
Selectivity relative to preceding peaks	✓	✓
Baseline noise	✓	✓
Signal-to-noise ratio	✓	✓
Baseline drift	✓	✓
Peak purity based on spectral evaluation (only with diode-array detector and HPLC^{3D} ChemStation or ^{3D}CE ChemStation)	✓	✓
For a series of consecutive runs		
Precision of peak retention/migration time	✓	✓
Precision of peak area,[a] heights or amounts	✓	✓
(linearity)	✓	✓

[a] Note that in CE, corrected peak area (area divided by migration time) is used for quantitative analysis

2.10 AUTOMATION OF UNIT PROCESSES

A problem that is easiest solved by automation of the constituent unit processes is that with a relatively simple analytical methodology, e.g. X-ray fluorescence spectrometry, which involves a minimum of sample treatment, applied to the continuous monitoring and control of a process stream. However, the problems that are currently solved by automated methods are many and varied and are chiefly encountered in the quality control of industrial processes, clinical analysis and environmental monitoring, areas in which large numbers of similar analyses have to be carried out (automated repetitive analyses) or continuous monitoring of analytes is necessary, in both cases with a minimum of human involvement. Such automated methods are therefore considerably faster than the corresponding manual methods and are frequently more reliable in the absence of subjective judgements by human operators. They also have the facility to be operated in situations where operators would find it either difficult or impossible to work.

2.10.1 Automation of Repetitive Analysis

Automated repetitive analysis consists essentially of the same unit processes as the corresponding manual method. Individual unit processes can be automated, e.g. sampling with an automatic pipette or solid-phase extraction prior to HPLC (Varian

Unit Processes of Analytical Procedures

9200 Prospekt system). Alternatively, the analytical method can be automated at all stages from sampling at a turntable onwards as pioneered by Technicon with their AutoAnalyser and subsequently by other organisations for the automated analysis of a range of inorganic and organic analytes in complex matrices. By the use of computer-controlled robots it is now possible to automate fully many of these analytical methods. This is further discussed in Sections 2.10.3, 2.10.4 and 2.10.5.

Figure 18 illustrates how such an autoanalyser operates for the automation of these unit processes of an analytical method involving a colorimetric/photometric measurement step. An automatic pipette extracts a specified volume of a liquid sample from one of a series of cups on a turntable. This sample is mixed with suitable reagents in appropriate proportions and driven on through the subsequent unit processes by a peristaltic proportioning pump. Air bubbles are incorporated as shown and at regular intervals to prevent diffusion and cross contamination between successive samples. The mixing coil allows for complete mixing of reagents and sample prior to the separator unit removing interfering species. Prior to colorimetric/photometric determination, the colour reagent is added to the stream and allowed to react with the analyte for a sufficient time period in the reaction coil, thermostated to a preset temperature. The coloured solution then enters the flow cell of the on-line detector via a 'debubbler' which removes the air. Such autoanalysers can process up to 150 samples per hour.

Figure 19 illustrates the automated determination of magnesium in urine utilising its chelation with o,o'-dihydroxyazobenzene (DAB) and subsequent fluorimetric detection.[42] The final reagent concentrations used were ethylenediamine 0.28 mol dm^{-3}, DAB 0.07 mmol dm^{-3}, HCl 0.14 mol dm^{-3} and KCl 0.10 mol dm^{-3}

Modern automated repetitive analysis in clinical biochemistry for organic analytes such as glucose, urea, creatinine and bilirubin and inorganic analytes such as potassium, sodium and chloride utilises alternative instrumentation to segmented flow autoanalysers. Figure 20 illustrates how samples are autopipetted together with appropriate reagents into reaction discs where absorbances at appropriate wavelengths are measured by a multi-wavelength photometer. Inorganic analytes are determined by ion-selective electrodes (Boehringer Mannheim/Hitachi 911 system). Serum glucose is a highly requested analyte in clinical biochemistry since it is diagnostic of uncontrolled diabetes mellitus and hyperactivity of endocrine glands (high concentration of glucose), and insulin overdose and liver disease (low concentration of glucose). In the past, glucose was determined using chemical reactions which were relatively time consuming and involved relatively high temperatures for operation in automated segmented flow autoanalysers. The aldehyde group in the open ring form of glucose could be condensed with o-toluidine to give a coloured product with a λ_{max} of 630 nm following 8 min boiling. Alternatively, this aldehyde group could reduce yellow Fe(CN)$_6^{3-}$ to the colourless Fe(CN)$_6^{4-}$ following heating at 95°C. The decrease in absorbance due to Fe(CN)$_6^{3-}$ was found proportional to the glucose concentration.

Nowadays, enzymatic methods are more commonly used for serum glucose

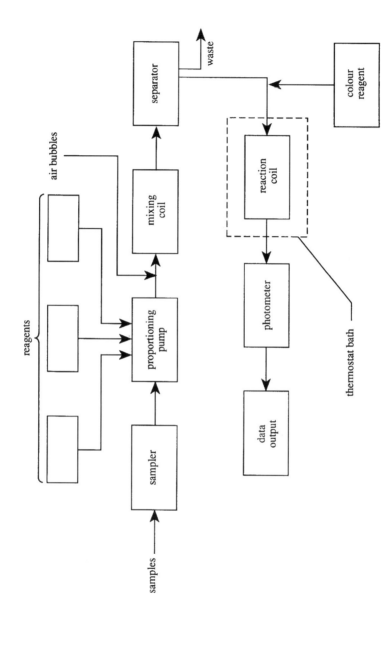

Figure 18 Schematic diagram showing operation of an autoanalyser incorporating air bubbles to prevent mixing between samples. Reproduced by permission of Chapman & Hall from F. W. Fifield and D. Kealey, *Principles and Practice of Analytical Chemistry*, 3rd edn, Blackie, Glasgow, 1990

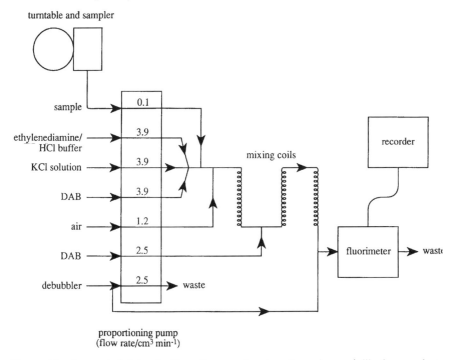

Figure 19 Automated determination of magnesium in urine using o,o'-dihydroxyazobenzene (DAB). Reproduced by permission of Chapman & Hall from F. W. Fifield and D. Kealey, *Principles and Practice of Analytical Chemistry*, 3rd edn, Blackie, Glasgow, 1990

determination using automated instruments such as the Boehringer Mannheim/ Hitachi 91 system (Figure 20). The hexokinase enzymatic method is recognised as highly specific, rapid, accurate and with the ability to measure low concentrations of serum glucose. Glucose is first phosphorylated by adenosine triphosphate (ATP) in a reaction catalysed by hexokinase (HK) to produce glucose-6-phosphate (G6P) and ADP. G6P is then oxidised to 6-phosphogluconate (6PG) by nicotinamide adenine dinucleotide (NAD), in the presence of catalyst glucose-6-phosphate dehydrogenase (G6PDH). The product of NAD reduction, i.e. NADH, can be monitored at 340 nm with the absorbance at this wavelength being directly proportional to the serum glucose concentration.

$$\text{glucose} + \text{ATP} \xrightarrow{\text{HK}} \text{G6P} + \text{ADP}$$
$$\text{G6P} + \text{NAD} \xrightarrow{\text{G6PDH}} \text{6PG} + \text{NADH}$$

Using the above-mentioned 911 instrumentation, expected values in the range 2.5–7.9 mmol dm^{-3} with technical limits of 0.5–24.5 mmol dm^{-3} are used and samples of 5 μl are subjected to the hexokinase enzymatic method. Any result below 0.5 mmol dm^{-3} is repeated with an increase in sample volume to 10 μl and any

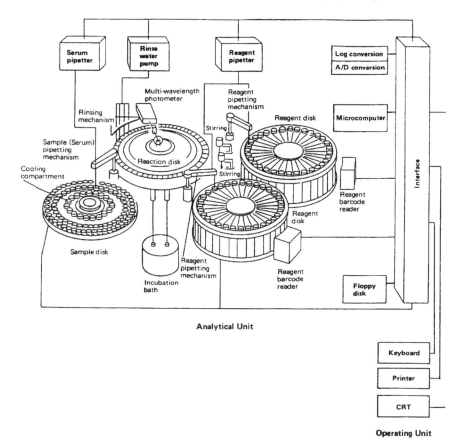

Figure 20 The Boehringer Mannheim/Hitachi 911 system as an example of modern automated repetitive analysis in clinical biochemistry. Reproduced by permission of Boehringer Mannheim Diagnostics

result above 24.5 mmol dm^{-3} is repeated with a decrease in sample volume to 2 μl. A control sample is assayed after every 50 samples and has a concentration of 5.2 mmol dm^{-3}. The time taken for the assay is 4 min.

A relatively recent development in automated analysis has been in the introduction of flow injection analysis in which a discrete volume of a liquid sample is injected into a carrier stream which already contains the reagents necessary for colour development when a colorimeter is used as detector. It is not necessary that this on-line colour-producing reaction is complete by the time the detector is reached, only that the analyte derivative is reproducibly dispersed by the time it reaches the detector and that it reaches the colorimeter in detectable concentrations. Dispersion is caused by both convective transport leading to a parabolic profile in the direction of flow and diffusional transport leading to primarily a radial profile in directions perpendicular to the flow. The extent of dispersion can be mathematically described as follows:

$$D = \frac{C_0}{C_{max}} \qquad (19)$$

where D is the dispersion coefficient, C_0 is the initial concentration of the analyte and C_{max} is the maximum concentration at the detector. Values of D in the ranges < 3, 3–10 and 10–15 represent limited, medium and large dispersions, respectively. Flow injection analysis is therefore a relatively simple analytical technique with sample throughputs which may exceed 100 per hour and has found particular application in routine monitoring of the aqueous environment and industrial process streams.

2.10.2 Continuous On-Line Monitoring

Continuous on-line monitoring is increasingly being used for the determination of a variety of analytes in essentially unattended situations such as on a river bank for the determination of soluble chemical entities such as Al^{3+}, PO_4^{3-}, SiO_2 and N_2H_4. The Hach Company (Loveland, CO, USA), for example, has patented Series 5000 Analyzers for the semi-continuous on-line monitoring of such analytes. The Aluminium Analyzer (Figure 21) uses chelation with Eriochrome Cyanine R (**XI**) to produce an Al^{3+} chelate absorbing at 540 mm to measure Al^{3+} concentration continuously in the aqueous environment over the concentration range 0–300 µg d-m^{-3} with an accuracy of $\pm 10\%$. Prior to colorimetric determination, a masking reagent is added to the sample to eliminate iron interference and buffer reagent is also added to adjust the pH to 6.00 ± 0.3. A blank is determined using addition of an EDTA reagent which chelates any Al^{3+} in the sample and prevents its reaction with (**XI**). The value of this blank, which represents 0 µg dm^{-3} Al^{3+}, is stored in the microprocessor and subtracted during subsequent sample measurements. An Al^{3+} standard solution (250 µg dm^{-3}) is used for autocalibration. The results can be corrected automatically for known fluoride concentrations following its independent determination and entry through the keyboard. The reagents are pressurised and dispensing is controlled by microprocessor-activated solenoid valves with no pumps

(**XI**)

Figure 21 A Hach Series 5000 continuous on-line monitoring analyser. The illustration shows the reagents and standard solutions in the lower chamber of the instrument. The control module is seen top right. The colorimeter is shown top left, below the flow display and the control valves. Reproduced by permission of Hach Company, Colorado, USA

to maintain—unattended operation is possible for 1 month. The Series 5000 Analyzer provides a digital display of concentration and it is possible to send the data from the Analyzer to an external computer, permitting storage of the Al^{3+} concentration data and external control of the instrument's programming and operation on the river bank.

2.10.3 Laboratory Robotics

The development of reliable robots together with sophisticated, programmable computer controllers has begun to make an impact on laboratory practice in analytical laboratories. For example, in the pharmaceutical industry with the imposition of more stringent regulatory demands, there has been a notable increase in the number of samples that have to be analysed. As a result, analytical and drug metabolism laboratories have been turned into 'bottlenecks' in the development of new drugs. Robotics, not surprisingly, is having a major influence in these laboratories with the

Unit Processes of Analytical Procedures 77

following advantages:

(a) sustained, prolonged and efficient operation;
(b) release of skilled operators for other duties;
(c) saving in staff costs;
(d) increased accuracy and precision;
(e) reduced human contact with hazardous and/or infective materials;
(f) improved product quality.

In these analytical and drug metabolism laboratories, the use of intelligent autosamplers, on-line data capture, data reduction and laboratory information management systems has improved their productivity. The overall productivity, however, is still limited to the number of samples that can be manually prepared by the analyst. Cost-effective, reliable laboratory robots are used increasingly to perform this and other laboratory unit operations (LUOs). In addition, robots are reprogrammable mechanical manipulators and their workstation is easily reconfigured, unlike dedicated equipment that can be made redundant by a change in sample type or protocol.

2.10.4 Application of Robotics to Dissolution Tests

Dissolution tests monitor *in vitro* the drug release rates of tablets and other solid dosage forms into a suitable medium in order to meet clinical and legislative requirements. There are several steps:

(a) preparation of the dissolution apparatus which includes dispensing the dissolution medium (H_2O or dilute HCl);
(b) introduction of the dosage form;
(c) removing and filtering aliquots of the dissolution medium at timed intervals and determination of the drug content of solutions by UV spectrophotometry;
(d) data processing and presentation.

At the end of this test cycle, the dissolution apparatus is cleaned and prepared for further tests.

Robotic systems are now used in dissolution tests by employing four specially designed hands for filling the dissolution apparatus, sample introduction, sampling and cleaning the apparatus. The robot can perform up to 12 fully unattended dissolution tests and can pay for itself in a year. A comparison of a robotic system and a manual method is given in Table 17, with the results expressed as a percentage of the drug released after 45 min.

2.10.5 Application of Robotics to Drug Determination in Biological Fluids

In a typical example, the robot does the manual operations necessary to extract the drug from the biological fluid and prepare it for direct analysis by HPLC. The plasma is sampled and an internal standard is added together with buffer and organic

Table 17 Comparison of manual and robotic methods for drug dissolution tests

Parameter	Robotic method	Manual method
No. of dissolutions N	72	36
Mean (%)	96.6	95.4
Coefficient of variation (%)	1.1	2.7
Range (%)	92.4–99.3	92.1–100.4

solvent, followed by a series of LUOs such as liquid–liquid extraction, centrifugation, organic layer separation, evaporation, dissolving residue by vortex mixing and transfer to autosampler vials. This is followed by automated HPLC with autosampler injection of the sample on to the HPLC column under control of a computing integrator which calculates, stores the results and prints a summary report after all samples have been processed. Such a robotic system can analyse 150 samples per day as compared with 80 per day using a manual method.

Chapter 3

Selected Analytical Problems Involving Inorganic Analytes which Contain Elements from Groups IA–VIIIA and the Lanthanides

3.1 DETERMINATION OF SODIUM AND POTASSIUM IN MINERAL WATER BY FLAME EMISSION SPECTROMETRY

3.1.1 Summary

Alkali metals such as Na and K are readily determined directly in solutions such as mineral water and biological fluids by flame emission spectrometry. The overall analytical method is relatively rapid and inexpensive with coefficients of variation in the range 1–4%. The technique itself can suffer from spectral interferences and self-absorption. Its main rival for the determination of the alkali metals and some alkaline earths such as Ca, when instrumental costs are considered, would be potentiometric measurement with appropriate ion-selective electrodes.

3.1.2 Introduction

Atomic emission spectrometry employing a flame (otherwise known as flame emission spectrometry or flame photometry) has found widespread application in elemental analysis. Elements emit characteristic spectra upon being aspirated into flames of H_2–O_2, C_2H_2–O_2 and C_2H_2–N_2O, which have temperatures in the range 2000–3000 K. A summary of flame emission data for the determination of Na, K, Li and Ca is given in Table 18. Because of its convenience, speed and relative freedom from interferences, flame emission spectrometry has become the method of choice

Table 18 Summary of flame emission data for the determination of Na, K, Li and Ca[42]

Element	Spectral line (nm)	Electronic transition	Flame	Detection limit (ppm)
Na	589	$3p \to 3s$	H_2-O_2	0.001
K	766	$4p \to 4s$	$C_2H_2-O_2$	0.001
Li	671	$2p \to 2s$	$C_2H_2-N_2O$	0.0001
Ca	423		$C_2H_2-N_2O$	0.001

for these relatively volatile elements with low excitation energies, which are otherwise difficult to determine. The technique has also been applied, with various degrees of success, to the determination of perhaps half the elements in the Periodic Table. Fully automated flame photometers (e.g. Instrumentation Laboratory IL 943) are used in clinical biochemistry laboratories for the determination of Na, K and Li in biological fluids. Only 20 μl of sample are required and with the use of a diluent, a caesium internal standard, all three analytes can be determined with coefficients of variation of 1–2.5% at the press of a single button.

3.1.3 Procedure

A radiation buffer which is a saturated solution of $CaCl_2$, KCl and $MgCl_2$, prepared in that order for the determination of Na in mineral water, is added to both standards and sample to minimise the effect of matrix species on the analyte signal. A radiation buffer for the determination of K is a saturated solution of NaCl, $CaCl_2$ and $MgCl_2$, prepared in that order. Standards in the 0–10 ppm concentration range for each analyte would be prepared in 100-cm^3 flasks with 5 cm^3 of the appropriate radiation buffer added in each case. Calibration plots of emission intensity corrected for the background signal would be plotted against concentration for each analyte. The mineral water sample, appropriately prepared by any necessary dilution and addition of radiation buffer, would then be aspirated into the flame and the concentration of each analyte determined separately by reference to the previously prepared calibration plots.

3.2 DETERMINATION OF WATER HARDNESS, i.e. TOTAL CALCIUM AND MAGNESIUM, BY EDTA TITRATION

3.2.1 Summary

Total Ca and Mg (i.e. water hardness) can be rapidly and particularly inexpensively determined by EDTA titration using Erichrome Black T as an indicator. This chemical method of analysis is very precise with coefficients of variation generally less than 1%. Spectroscopic methods of analysis such as atomic absorption spectrometry have effectively replaced EDTA titrations for the determination of Ca and Mg in those laboratories that can afford their purchase and upkeep. A worked calculation on water hardness estimation is provided.

3.2.2 Introduction

The determination of water hardness is a useful analytical test that provides a measure of the quality of water for household and industrial uses. The test is important to industry because hard water, upon being heated, precipitates $CaCO_3$, which then clogs boilers and pipes. Water hardness can be determined by an EDTA titration after the sample has been buffered to pH 10, which prevents competition of H_3O^+ for the EDTA anion. Higher pH values would cause precipitation of $CaCO_3$ or $Mg(OH)_2$. Magnesium, which forms the least stable EDTA complex of all of the common multivalent cations in typical water samples, is not titrated until enough reagent has been added to complex all of the other cations in the sample. Therefore, a magnesium ion indicator such as Eriochrome Black T (**XII**) can serve as an indicator in water-hardness titrations. Often a small concentration of the EDTA complex of magnesium is incorporated in the buffer or in the titrant to ensure the presence of sufficient Mg^{2+} for satisfactory indicator action. On the addition of $Na_2Mg(EDTA)$, which does not affect the stoichiometry of the titration reaction since the salt contains equimolar quantities of Mg and EDTA, the EDTA is exchanged from Mg^{2+} to Ca^{2+} since the formation constant of $[CaEDTA]^{2-}$ is about two orders of magnitude higher than that of $[MgEDTA]^{2-}$, i.e. $K_{CaEDTA^{2-}} = 5 \times 10^{10}$ and $K_{MgEDTA^{2-}} = 4.9 \times 10^{8}$:

$$[MgEDTA]^{2-} + Ca^{2+} \rightarrow Mg^{2+} + [CaEDTA]^{2-}$$

(**XII**)

The Mg then chelates to the indicator (**XII**) (the formation constant of the Ca–Erichrome Black T complex is approximately one fortieth of that of the Mg complex with Erichrome Black T):

$$[Ca\ Erichrome\ Black\ T]^- + Mg^{2+} \rightarrow Ca^{2+} + [Mg\ Erichrome\ Black\ T]^-\ red$$

As EDTA titrant is added, it first binds to all the Ca^{2+}:

$$Ca^{2+} + H_xEDTA^{x-4} \rightarrow [CaEDTA]^{2-} + xH^+$$

until at the end-point EDTA displaces the less strongly bound Erichrome Black T from Mg^{2+}, resulting in the end-point colour change from red to blue:

$$[\text{Mg Erichrome Black T}]^- + \text{H}_x\text{EDTA}^{x-4} \rightarrow$$
red

$$[\text{MgEDTA}]^{2-} + [\text{H Erichrome Black T}]^{2-} + (x-1)\text{H}^+$$
blue

3.2.3 Procedure

(a) Test kits for determining the hardness of household water are available commercially. They consist of a vessel calibrated to contain a known volume of water, a measuring scoop to deliver an appropriate amount of a solid buffer mixture, an indicator solution and a bottle of standard EDTA which is equipped with a medicine dropper. The volume of standard reagent consumed is obtained by counting drops to the end-point. The concentration of the EDTA solution is ordinarily such that one drop corresponds to one grain (ca 0.065 g) of calcium carbonate per gallon of water. Although imprecise by common laboratory standards, such determinations are rapidly performed and are satisfactory for monitoring water treatment processes.

(b) A laboratory assessment of water hardness would be made as follows. Acidify 100-cm^3 aliquots of the sample with a few drops of HCl and boil gently for a few minutes to eliminate CO_2. Cool, add 3–4 drops of methyl red and neutralise with 0.1 mol dm^{-3} NaOH. Introduce 2 cm^3 of pH 10 buffer, 3–4 drops of Erichrome Black T, 1–2 cm^3 of 0.1 mol dm^{-3} [MgEDTA]$^{2-}$ and titrate with standard 0.01 mol dm^{-3} Na$_2$H$_2$EDTA to a colour change from red to pure blue. The result is expressed in terms of milligrams of $CaCO_3$ per litre of water.

3.2.4 Calculation

A 50-cm^3 water required 21.76 cm^3 of 0.02 mol dm^{-3} EDTA to titrate water hardness. What was the hardness in mg $CaCO_3$ dm^{-3}?

$$\text{No. of moles EDTA} = \frac{21.76}{1000} \times 0.02$$

$$\text{No. of moles of Ca} + \text{Mg} = \frac{21.76}{1000} \times 0.02$$

$$\text{Weight expressed as CaCO}_3 = \frac{21.76}{1000} \times 0.02 \times 100$$
$$= 0.0435 \text{ g}$$

$$\text{Weight of CaCO}_3 \text{ expressed as mg per dm}^3 \text{ of water sample} = 0.0435 \times 20 \times 1000$$
$$= 870 \text{ mg dm}^{-3}$$

3.3 GRAVIMETRIC DETERMINATION OF CERIUM IN ORE

3.3.1 Summary

Cerium could be conveniently and very inexpensively estimated gravimetrically on-site at an appropriate mine by precipitation as $Ce(IO_3)_4$, removal of inferences, ignition of precipitate and weighing of the residue as CeO_2. Although the method is very precise with a coefficient of variation ideally less than 1%, it is time consuming and the accuracy will depend on the analytical scientist carrying out the procedure very carefully and making very accurate weighings. Cerium, other lanthanides and elements present in the ore could be rapidly identified and determined simultaneously by resort to inductively coupled plasma mass spectrometry, only available at well funded university, government and industrial laboratories.

3.3.2 Introduction

The lanthanides, including La and Y, were originally known as the rare earths, from their occurrence in oxide (or in old usage, earth) mixtures. They are not rare elements and their absolute abundances are relatively high. Thus even the scarcest, thulium (Tm), is as common as bismuth (Bi) and more common than As, Cd, Hg or Se. The major source is monazite, a heavy, dark sand of variable composition. Monazite is essentially a lanthanide orthophosphate but may contain up to 30% thorium (Th). Lanthanum (La), cerium (Ce), praseodymium (Pr) and neodymium (Nd) usually account for ca 90% of the lanthanide content of these minerals, with yttrium (Y) and the heavier elements accounting for the rest.

3.3.3 Procedure

Cerium in its +4 oxidation state may be separated from interferences by either (a) solvent extraction of Ce^{4+}, Zr^{4+}, Th^{4+} and Pu^{4+} from nitric acid solution with tributyl phosphate dissolved in kerosene or another inert solvent, thus separating these +4 ions from other +3 lanthanide ions or (b) precipitation of Ce^{4+}, Th^{4+}, Ti^{4+} and Zr^{4+} as iodates, e.g. $Ce(IO_3)_4$, insoluble in 6 mol dm^{-3} HNO_3, thus separating these +4 ions from the other +3 lanthanide ions. In both cases the resulting determination technique either must be able to discriminate Ce^{4+} from the other tetravalent species or the latter must be masked or removed prior to Ce^{4+} determination.

Using the gravimetric method (b), Ce^{4+}, as precipitated $Ce(IO_3)_4$, is ignited and weighed as cerium(IV) oxide (CeO_2) after removal of interferences such as coprecipitated $Th(IO_3)_4$, which can be dissolved and precipitated as the oxalate.

The ideal solution for the determination of Ce^{4+} would be a volume of 50 cm^3 with all metallic elements present as nitrates and the Ce^{4+} content should not exceed 0.10 g (2000 ppm). This solution is treated with half its volume of concentrated HNO_3 with 0.5 g of $KBrO_3$ added to oxidise Ce to the +4 oxidation state. When the latter dissolves, 10–15 times the theoretical quantity of KIO_3 in HNO_3 solution (prepared by dissolving 50 g of KIO_3 in 167 cm^3 of concentrated HNO_3 and diluting to 500 cm^3) is added slowly and, with constant stirring, the precipitated $Ce(IO_3)_4$ is allowed to settle. When cold, the precipitate is filtered through a fine filter-paper (e.g. Whatman No. 42 or 542), allowed to drain, rinsed once and then washed back into the beaker in which precipitation took place by means of a solution of 0.8 g of KIO_3 and 5 cm^3 of concentrated HNO_3 in 100 cm^3. After thorough mixing, the precipitate is collected on the same paper and then washed back into the beaker with hot water. This is boiled and treated with concentrated HNO_3 dropwise until the precipitate just dissolves (20–25 cm^3 of HNO_3 are required per 0.1 g of cerium). A 0.25-g amount of $KBrO_3$ and as much KIO_3–nitric acid solution as before are then added. When cold, $Ce(IO_3)_4$ is collected and washed prior to its treatment in the same beaker with 5–8 g of oxalic acid and 50 cm^3 of water and heated to boiling. After all the iodine has been expelled, the precipitate is set aside for several hours, then filtered, washed with cold water, dried and ignited at 500–600°C to constant weight in a Pt crucible. The resulting solid is weighed as CeO_2.

3.4 SPECTROPHOTOMETRIC DETERMINATION OF TITANIUM IN ROCK

3.4.1 Summary

Titanium exists in rocks as a minor constituent in the range 0.1–1%. It can be conveniently assayed in this matrix by dissolution with a mixture of HF, HNO_3 and H_2SO_4, formation of the intensely coloured complex ion with H_2O_2 and its measurement by visible spectrophotometry at 410 nm. Such an overall analytical method is relatively rapid and inexpensive and could be carried out on-site. The coefficient of variation of the instrumental technique itself would be expected to be in the range 0.5–5% and possible interferences such as molybdenum and vanadium could be dealt with by the solution of simultaneous equations as illustrated in the calculation (Section 3.4.4).

3.4.2 Introduction

Titanium is relatively abundant in the earth's crust (0.6%), its main ores being ilmenite ($FeTiO_3$) and rutile, one of the several crystalline varieties of TiO_2. Titanium is lighter than other metals of similar mechanical and thermal properties and is unusually resistant to corrosion. It is used in turbine engines and industrial chemical, aircraft and marine equipment. It is unattacked by dilute acids and bases. For the relatively small concentrations of Ti usually occurring in rocks, its determination at 410 nm following formation of the intense yellow colour generated from Ti(IV) in acidic solution together with H_2O_2 has been widely used (Figure 22). The resulting complex ion has been described as $[TiO(SO_4)_2]^{2-}$, $[Ti(H_2O_2)]^{4+}$ or $[Ti(O_2)(OH)]^+$. V, Mo and Cr, under certain conditions, also form coloured species with H_2O_2 but usually these are not present in sufficient amounts in rocks to interfere with the Ti determination. If they did, as in Figure 22, then additivity of the absorbances at 330,

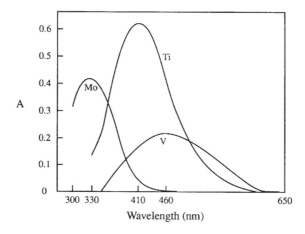

Metal	Absorbance (A)		
	330 nm	410 nm	460 nm
Mo	0.416	0.048	0.002
Ti	0.130	0.608	0.410
V	0.000	0.148	0.200

Figure 22 UV–visible spectral behaviour and data for Mo, Ti and V complexes with H_2O_2 in acidic solution. Reprinted with permission from A. Weissler, *Ind. Eng. Chem., Anal. Ed.*, 1945, **17**, 695. Copyright 1945 American Chemical Society

410 and 460 nm and subsequent solution of simultaneous equation resulting from application of the Beer–Lambert law could be used for the simultaneous determination of Mo, Ti and V by visible spectrophotometry.[50] Other elements such as Fe, Ni, Co and Cu can also interfere because their solutions are coloured. The interference of these elements can be overcome by placing the appropriate sample or standard, not treated with H_2O_2, in the reference cell of the spectrophotometer. Traces of fluoride have a marked bleaching effect on the colour of the peroxo complex of Ti. Fluoride must therefore be removed by repeated evaporation with concentrated H_2SO_4 to expel all HF. The bleaching effect of relatively large concentrations of phosphates or alkali metal salts can be overcome by adding like amounts to the standards.

3.4.3 Procedure

A standard titanium solution can be prepared as follows. A 3.68-g amount of AR potassium titanyl oxalate [$K_2TiO(C_2O_4)_2 \cdot 2H_2O$] is weighed into a Kjeldahl flask, 8 g of $(NH_4)_2SO_4$ and 100 cm^3 of concentrated H_2SO_4 are added and the mixture is gradually heated to boiling. Boiling is carried out for 10 min prior to cooling and dilution with dilute H_2SO_4 to 1 dm^3 in a graduated flask. This solution should have a pH of ca 0 and is 500 ppm with respect to Ti. Standards in the range 10–100 ppm Ti can be prepared in 100-cm^3 graduated flasks after reaction of appropriate volumes of the 500 ppm stock solution with 10 cm^3 of 3% H_2O_2 and making up to the mark with dilute H_2SO_4 so as to maintain a pH of ca 0 in the standard solutions. A calibration plot of absorbance at 410 nm vs ppm Ti can then be constructed.

A 1-g amount of the powdered rock sample, which could contain 0.1–1% Ti, can be decomposed in a Pt dish with a mixture of HF, HNO_3 and H_2SO_4 by heating to evaporation of fumes of SO_3, cooling, addition of necessary H_2SO_4 and further evaporation until all HF has been removed. The sample solution is then made up in H_2SO_4 to a final pH of ca 0. A 10-cm^3 volume of 3% H_2O_2 is added and the resulting solution is made up to 100 cm^3 with H_2SO_4 to maintain a pH of ca 0 and to give a solution which will have a concentration of Ti in the range 10–100 ppm. Its absorbance at 410 nm can then be referred to the calibration plot in order to calculate the percentage of Ti in the original rock sample. Due account must be taken of potential interferences, as discussed earlier.

3.4.4 Calculation

A 1-g amount of a powdered rock sample which contains Ti, Mo and V is dissolved using HF, HNO_3 and H_2SO_4, reacted with H_2O_2 and made up to 100 cm^3 with H_2SO_4 as illustrated in the procedure (Section 3.4.3). The absorbance of this solution at 330 nm is 0.2, at 410 nm is 1.0 and at 460 nm it is 0.4. Use the data shown in Figure 22 where each metal at a concentration of 4 mg in 100 cm^3 of H_2SO_4 gives the individual absorbances shown at 330, 410 and 460 nm to calculate the percentage of Ti in the original rock sample. Assume the spectrophotometric cells have a path length of 1 cm.

Expressing concentrations in mg per 100 cm^3 and the path length as 1 cm, the molar absorptivities or extinction coefficients (ε) for the three metals at the three wavelengths can be calculated from the Beer–Lambert law as follows:

$$\varepsilon_{Mo}^{330} = \frac{0.416}{4}; \quad \varepsilon_{Ti}^{330} = \frac{0.130}{4}; \quad \varepsilon_{V}^{330} = 0$$

$$\varepsilon_{Mo}^{410} = \frac{0.048}{4}; \quad \varepsilon_{Ti}^{410} = \frac{0.608}{4}; \quad \varepsilon_{V}^{410} = \frac{0.148}{4}$$

$$\varepsilon_{Mo}^{460} = \frac{0.002}{4}; \quad \varepsilon_{Ti}^{460} = \frac{0.410}{4}; \quad \varepsilon_{V}^{460} = \frac{0.200}{4}$$

Using additivity of absorbances for the unknown solution measured at 330, 410 and 460 nm, respectively, where C_{Mo}, C_{Ti} and C_V represents the concentrations of Mo, Ti and V in mg per 100 cm^3:

$$0.2 = \frac{0.416}{4} \times C_{Mo} \times 1 + \frac{0.130}{4} \times C_{Ti} \times 1 \tag{20}$$

$$1.0 = \frac{0.048}{4} \times C_{Mo} \times 1 + \frac{0.608}{4} \times C_{Ti} \times 1 + \frac{0.148}{4} \times C_V \times 1 \tag{21}$$

$$0.4 = \frac{0.002}{4} \times C_{Mo} \times 1 + \frac{0.410}{4} \times C_{Ti} \times 1 + \frac{0.200}{4} \times C_V \times 1 \tag{22}$$

These three equations can then be solved for three unknowns. Solving for C_{Ti} as asked for in the calculation:
from equation 20:

$$C_{Mo} = \frac{0.2 - 0.0325 C_{Ti}}{0.1040} \qquad (23)$$

from equation 22:

$$C_V = \frac{0.4 - 0.0005 C_{Mo} - 0.1025 C_{Ti}}{0.05} \qquad (24)$$

Substitute equations 23 and 24 in equation 21:

$$1.0 = 0.0120 \left(\frac{0.2 - 0.0325 C_{Ti}}{0.1040} \right) + 0.1520 C_{Ti} + 0.0370 \left(\frac{0.4 - 0.0005 C_{Mo} - 0.1025 C_{Ti}}{0.05} \right) \qquad (25)$$

Substitute the value for C_{Mo} (i.e. from equation 23) in equation 25:

$$1.0 = 0.023 - 0.0037 C_{Ti} + 0.1520 C_{Ti} + 0.0370(8 - 0.019 + 0.003 C_{Ti} - 2.0500 C_{Ti})$$

Rearranging:

$$0.0874 C_{Ti} = 0.7376$$

$$C_{Ti} = 8.44 \text{ mg per } 100 \text{ cm}^3$$

There are therefore 8.44 mg of Ti in 1 g of the original rock sample, i.e. 0.844% w/w.

3.5 DETERMINATION OF VANADIUM(V) IN SEA WATER BY ADSORPTIVE STRIPPING VOLTAMMETRY

3.5.1 Summary

Adsorptive stripping voltammetry can be used to determine vanadium(V) directly in sea water, following destruction of naturally occurring surfactants and chelation of V(V) with catechol. Other potential interferences such as Sb and U, which also form adsorbable, reducible chelates, are not reduced at the same potential and/or give relatively low non-interfering signals. The method is rapid and uses relatively inexpensive instrumentation which is also available for usage in the field and in mobile laboratories. Expected coefficients of variation are 5% or less.

3.5.2 Introduction

The concentration of dissolved vanadium in sea water lies in the range $(2-3) \times 10^{-8}$ mol dm^{-3}. The predominant oxidation state is V(V), the concentration of which is normally determined spectrophotometrically after preconcentration methods such as coprecipitation and solvent extraction. Voltammetric techniques such as polarography of V(V) or cathodic stripping voltammetry of the Hg(I) salt of the V(V) anion are insufficiently sensitive to determine vanadium in sea water, whereas adsorptive stripping voltammetry of the catechol chelate with V(V) has an LOD of 0.1 nmol dm^{-3} vanadium under optimum analytical conditions and can thus be used to determine dissolved vanadium in sea water directly.[51]

3.5.3 Procedure

Van den Berg and Huang[51] have used adsorptive stripping voltammetry to determine V(V) directly in sea water using the following optimum electroanalytical conditions: catechol concentration 2×10^{-4} mol dm^{-3}, accumulation potential for the hanging mercury drop electrode -0.1 V, supporting electrolyte pH 6.9 and accumulation time 15 min in stirred solution. This resulted in an adsorptive stripping peak at ca -0.7 V with an LOD of 0.1 nmol dm^{-3}. There are many potential interferences since other metals such as Pb, U, Sb, Zn, Cd, Fe and Bi can also form adsorbable reducible chelates on the hanging mercury drop electrode. It was found that the V peak was preceded by that of U and followed by that of Sb. The sensitivity towards Sb was found to be low and the U peak at ca 0.18 V more positive than V was found not to interfere at common concentrations in sea water. Natural organic surfactants in

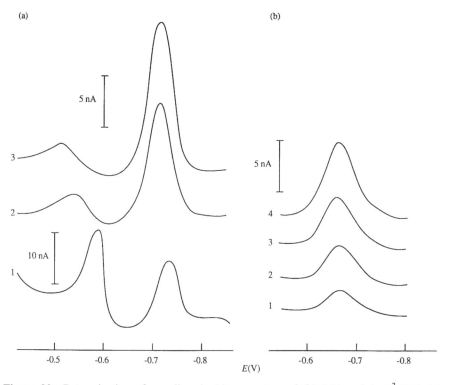

Figure 23 Determination of vanadium in (a) sea water and (b) 0.01 mol dm^{-3} KCl. (a) Collection time, 1 min. Scans: 1, LSCSV, scan rate 50 mV s^{-1}; 2, DPCSV, scan rate 5 mV s^{-1}, pulse rate 2 s^{-1}; 3, DPCSV, scan rate 1 mV s^{-1}, pulse rate 10 s^{-1}. (b) Collection time 2 min, DPCSV, scan rate 1 mV s^{-1}, pulse rate 10 s^{-1}. Scans: 1, 3×10^{-9} mol dm^{-3} vanadium; 2, 5×10^{-9} mol dm^{-3} vanadium; 3, 7×10^{-9} mol dm^{-3} vanadium; 4, 11×10^{-9} mol dm^{-3} vanadium. Reprinted with permission from Van den Berg and Huang, *Anal. Chem.*, 1984, **56**, 2383. Copyright 1984 American Chemical Society

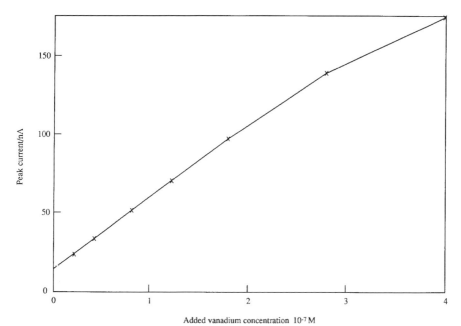

Figure 24 Peak height as a function of the vanadium concentration in sea water, measured by DPCSV (scan rate 10 mV s^{-1}, pulse rate 10 s^{-1}) after a 1 min collection. Initial vanadium concentration in the sample, 3×10^{-8} mol dm^{-3}. Reprinted with permission from Van den Berg and Huang, *Anal. Chem.*, 1984, **56**, 2383. Copyright 1984 American Chemical Society

sea water were found to interfere and have to be removed by UV irradiation prior to electroanalysis. The appropriate adsorptive stripping voltammetric traces are given in Figure 23 and the calibration plot is shown in Figure 24.

3.6 DETERMINATION OF TRACE CONCENTRATIONS OF MOLYBDENUM IN WATER SAMPLES BY SOLVENT EXTRACTION OF A NEUTRALLY CHARGED CHELATE FOLLOWED BY ATOMIC ABSORPTION SPECTROMETRY

3.6.1 Summary

Low ppb levels of Mo in water samples can be determined by concentration of Mo through solvent extraction of its neutrally charged chelate with 8-hydroxyquinoline into a relatively small volume of methyl isobutyl ketone followed by atomic absorption spectrometry (AAS) in either the flame or carbon furnace mode. This overall method gives low LODs for Mo and other metals[39] but is slow. The technique of atomic absorption spectrometry itself is probably the most popular tool for the trace determination of elements in a variety of matrices and covers some 70 elements in the Periodic Table. Coefficient of variation for the technique is in the range 0.5–2%.

3.6.2 Introduction

An FWPCA survey[52] has revealed Mo in 33% of samples from US rivers with a mean of observed levels at 68 ppb. The normal range of levels in sea water has been quoted as 9–11 ppb.[53] Flame AAS, with an acetylene–nitrous oxide flame and suppressing interferences with an excess of a refractory element (1000 ppm Al), has an LOD of 40 ppb for Mo. Mo can be determined by carbon furnace AAS with an LOD of 4 ppb for a 5-μl sample. Solvent extraction of a neutrally charged chelate is recommended for the determination of low Mo concentrations and for sea-water determinations.

3.6.3 Procedure

Solvent extraction/concentration of Mo using the chelating agent 8-hydroxyquinoline and the solvent methyl isobutyl ketone (MIBK) has been recommended.[54] A 100-cm^3 volume of the water sample is pipetted into a beaker, 5 cm^3 of 1% ascorbic acid as interference suppressor are added and the pH is adjusted to 2.2 ± 0.4. This sample is rinsed into a 150-cm^3 separating funnel and 5 cm^3 of the chelating agent solution (1% 8-hydroxyquinoline in MIBK) are added, prior to solvent extraction by shaking on a mechanical shaker for 15 min (this results in a concentration factor of ca 20). The organic phase is then separated and aspirated directly into the AAS flame. For determination by carbon furnace atomisation, solvent extraction is still recommended but a lower concentration factor than 20 can be used. Mo standards are prepared by spiking Mo-free samples of the same matrix (this is prepared by extracting all the Mo from a water sample as above prior to preparation of Mo standards and the blank) and solvent extracting by the above procedure. A blank is prepared by solvent extracting Mo-free samples of the same matrix (prepared as above).

3.7 DIFFERENTIAL THERMAL ANALYSIS (DTA) OF A COMPLEX MANGANESE COMPOUND, Mn(PH$_2$O$_2$)H$_2$O

3.7.1 Summary

DTA is one of a group of thermal methods of analysis which includes thermogravimetric analysis (TGA), differential scanning calorimetry and pyrolysis–gas chromatography. DTA, although not as widely used as other identification techniques such as infrared spectrometry and mass spectrometry, can provide complementary data for qualitative analysis.

3.7.2 Introduction

DTA is based upon the measurement of the temperature difference (ΔT) between the sample and an inert reference such as glass or Al$_2$O$_3$ as they are both subjected to the same heating programme. The temperature of the reference will thus rise at a steady rate determined by its specific heat and the programmed rate of heating. Similarly with the sample, except that when an exothermic or endothermic process occurs a peak or trough will be observed. The combination of exotherms and endotherms is unique to a particular sample. Thus the pattern of the DTA thermogram can be used as a fingerprint for qualitative analysis, while the area under a particular peak may be used for quantitative analysis. The technique can be widely applied to samples of very different types, e.g. minerals, inorganic compounds, pharmaceuticals, polymers, foodstuffs and biological specimens. Typical samples sizes are from 1 mg upwards allowing, where necessary, for measurements to be made on small samples. DTA may therefore be used effectively in a simple characterization or a purity estimate by studying the DTA characteristics of a particular sample in comparison with a standard.

3.7.3 Procedure

The complex manganese compound $Mn(PH_2O_2)H_2O$ is subjected to DTA and the resulting fingerprint is given in Figure 25.[42]

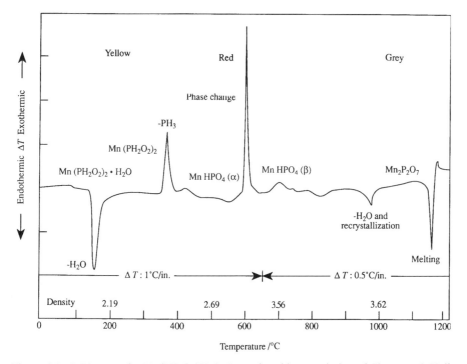

Figure 25 DTA curve for $Mn(PH_2O_2)H_2O$. Reproduced by permission of Chapman & Hall from F. W. Fifield and D. Kealey, *Principles and Practice of Analytical Chemistry*, 3rd edn, Blackie, Glasgow, 1990

3.8 DETERMINATION OF IRON IN AN ORE BY REDOX TITRATION WITH POTASSIUM PERMANGANATE

3.8.1 Summary

This relatively lengthy redox titrimetric procedure for the determination of iron in an ore could be carried out inexpensively on-site by a well trained analytical scientist with excellent accuracy and precision. The method is based on dissolution of the sample, prereduction of Fe and finally titration of Fe^{2+} with MnO_4^-. A worked calculation on Fe estimation is provided.

3.8.2 Introduction

The common ores of iron are hematite (Fe_2O_3), magnetite (Fe_3O_4) and limonite ($2Fe_2O_3.3H_2O$). The first step in the determination is (a) *dissolution of the sample*. Iron ores often decompose completely in concentrated HCl. The rate of attack by this reagent is increased by the presence of a small amount of $SnCl_2$, which probably acts by reducing sparingly soluble iron(III) oxides on the surface of the ore to more soluble iron(II) species. The tendency of iron(II) and iron(III) to form chloro complexes accounts for the effectiveness of HCl over HNO_3 or H_2SO_4 as a solvent for iron ores. Many iron ores contain silicates that may not be entirely decomposed by HCl. Incomplete decomposition is indicated by a dark residue which remains after prolonged treatment with the acid. A white residue of hydrated silica, which does not interfere in the subsequent determination, is indicative of complete decomposition.

(b) *Prereduction of iron* is the second step, since it is necessary to reduce all of the iron after dissolution to the $+2$ oxidation state prior to titration with the $KMnO_4$ oxidant. Excess $SnCl_2$ is frequently used for this purpose:

$$2Fe^{3+} + Sn^{2+} \rightarrow 2Fe^{2+} + Sn^{4+}$$

The remaining $SnCl_2$ is eliminated by the addition of excess $HgCl_2$:

$$Sn^{2+} + 2HgCl_2 \rightarrow Hg_2Cl_2(s) + Sn^{4+} + 2Cl^-$$

The slightly soluble Hg_2Cl_2 does not reduce $KMnO_4$, nor does the remaining $HgCl_2$ reoxidise Fe^{2+}. Care must be taken, however, to prevent the occurrence of the alternative reaction

$$Sn^{2+} + HgCl_2 \rightarrow Hg(l) + Sn^{4+} + 2Cl^-$$

since Hg reacts with $KMnO_4$ resulting in 'high' results. The formation of Hg, which is favoured by an appreciable excess of Sn^{2+}, is prevented by careful control of this excess and by the rapid addition of excess $HgCl_2$. A successful reduction is indicated by the appearance of a small amount of silky white precipitate after the addition of $HgCl_2$. Formation of a grey precipitate indicates the presence of elemental Hg. The absence of a precipitate indicates that an insufficient amount of $SnCl_2$ was used. In these latter two cases, the sample must be discarded.

(c) *Titration of* Fe^{2+} is the third step and involves the rapid reaction of Fe^{2+} with the permanganate anion, MnO_4^-:

$$MnO_4^- + 5Fe^{2+} + 8H^+ \rightarrow Mn^{2+} + 5Fe^{3+} + 4H_2O$$

The oxidation of Cl^-, which gives a positive titration error, is prevented by adding Zimmerman–Reinhardt solution, which contains $MnSO_4$, sulphuric acid and phosphoric acid. The Mn^{2+} ion shifts the equilibrium of the half-reaction

$$MnO_4^- + 8H^+ + 5e \rightleftharpoons Mn^{2+} + 4H_2O$$

to the left so that MnO_4^- is a less powerful oxidising agent with a lessened tendency to oxidise Cl^-. The phosphoric acid complexes Fe(III), thus increasing the tendency towards oxidation of Fe(II) to Fe(III), in preference to the oxidation of Cl^-. In addition, complexation of the yellow Fe^{3+} ion to form less highly coloured $FeHPO_4^+$ makes the end-point colour more readily visible.

3.8.3 Procedure

The ore is dried at 110°C for at least 3 h and then allowed to cool to room temperature in a desiccator. Samples of the order of 0.5 g are weighed into 500-cm^3 conical flasks and to each are added 10 cm^3 of concentrated HCl and ca 3 cm^3 of 0.25 mol dm^{-3} $SnCl_2$. Each flask is covered with a small watch-glass and heated at just below boiling until the samples have dissolved and undissolved solid, if any, is pure white. Another 1 or 2 cm^3 of $SnCl_2$ is used to eliminate any yellow colour that may develop as the solutions are heated. A blank of 10 cm^3 of concentrated HCl + 3 cm^3 of $SnCl_2$ is heated for the same amount of time. After the ore has been dissolved, the excess Sn^{2+} is removed by the dropwise addition of 0.02 mol dm^{-3} $KMnO_4$ until the solutions become faintly yellow. Dilution to ca 15 cm^3 is carried out and sufficient $KMnO_4$ solution added to impart a faint pink colour to the blank, then decolorisation is effected with one drop of the $SnCl_2$ solution. Each sample or blank should then be treated separately for steps (b) and (c), i.e. prereduction of iron and titration of Fe^{2+} to minimise air oxidation of Fe^{2+}. The sample solution is heated to near boiling and dropwise additions of 0.25 mol dm^{-3} $SnCl_2$ are made until the yellow colour just disappears, then two more drops are added. The solution is cooled to room temperature and 10 cm^3 of 5% $HgCl_2$ solution are added rapidly. A small amount of silky white Hg_2Cl_2 should precipitate; if not, the sample should be discarded. The blank solution should be treated with the $HgCl_2$ solution. Following addition of $HgCl_2$, 2–3 min should elapse before the addition of 25 cm^3 of Zimmermann–Reinhardt reagent and 300 cm^3 of water. Immediate titration with standard 0.02 mol dm^{-3} $KMnO_4$ should then be carried out to the first faint pink colour that persists for 15–20 s. The titrant volume should be corrected for the blank. The percentage of Fe can then be calculated for each sample.

3.8.4 Calculation

A 0.5349-g amount of a dried sample of an iron ore was dissolved in concentrated HCl and the iron was reduced to the +2 state by $SnCl_2$ prior to redox titration with 0.02759 mol dm^{-3} $KMnO_4$, of which 34.73 cm^3 were required to the end-point. Calculate the percentage of Fe in the sample.

$$\text{No. of moles of KMnO}_4 = \frac{34.73}{1000} \times 0.02759$$

$$\text{No. of moles of Fe}^{2+} = \frac{34.73}{1000} \times 0.02759 \times 5$$

$$\text{Mass of Fe}^{2+} = \frac{34.73}{1000} \times 0.02759 \times 5 \times 55.85 \text{ g}$$
$$= 0.2676 \text{ g}$$

$$\text{Fe}(\%) = \frac{0.2676}{0.5349} \times 100$$
$$= 50.02$$

3.9 DETERMINATION OF COBALT IN SOIL SAMPLES BY VISIBLE SPECTROPHOTOMETRY

3.9.1 Summary

Ppm concentrations of Co in soil samples can be estimated by ashing and dissolution of the sample, removal of Fe interference by extraction of $FeCl_4^-.H_3O^+$ into diethyl ether, reaction of Co with 1-nitroso-2-naphthol and determination of Co by visible spectrophotometry. Flame atomic absorption spectrometry, as applied to the ashed and dissolved sample, is faster and has greater selectivity. The visible spectrophotometric method could ideally be applied in the field or in a developing country where no atomic absorption spectrometer was available. A worked calculation on Co determination in a soil sample is given.

3.9.2 Introduction

Soils derived from granitic sources are often low in cobalt and this can affect the health of grazing animals, e.g. pining in sheep. Hence the determination of cobalt at ppm concentrations in soil samples is of significant agricultural importance.

3.9.3 Procedure and Calculation

A 5-g amount of a soil sample was heated for 16 h at 500°C in a platinum crucible in order to destroy organic matter. The residue was taken up in 100 cm^3 of concentrated

HCl and a 20-cm^3 aliquot of this solution was solvent extracted with 40 cm^3 of diethyl ether in a 100-cm^3 separating funnel. Fe was extracted into the organic layer as the ion pair $FeCl_4^-.H_3O^+$. The aqueous layer was then separated from the diethyl ether layer and the former was evaporated to almost dryness on a water-bath. Then 0.5 cm^3 of concentrated HCl + six drops of concentrated HNO$_3$ were added and the solution was made up to 15 cm^3 with distilled water, followed by boiling for 2 min. After cooling, 2 cm^3 of a 0.1% solution of 1-nitroso-2-naphthol plus 2 g of CH$_3$COONa were added and the solution was heated to 70°C. Five drops of phenolphthalein were added and then 10% KOH until the colour changed to pink. A further three drops of phenolphthalein were added and sufficient 0.5 mol dm^{-3} HCl to destroy the pink colour. The solution was then boiled for 2 min, cooled and diluted to 50 cm^3. This procedure results in the formation of a neutrally charged Co(III) chelate (**XIII**). 1-Nitroso-2-naphthol was one of the first selective organic reagents, discovered in 1885. It is selective for the determination of Co in the presence of Ni. Potential interferents include Bi(III), Cr(III), Hg(II), Sn(IV), Ti(III), W(VI), U(VI) and V(V).

(**XIII**)

The absorbance of the solution was then measured using a cell with path length 4 cm in order to facilitate the determination of particularly low Co concentrations. The Co concentration of this unknown was then determined by comparison with standards prepared as follows: 0.434 g of CoCl$_2$.6H$_2$O (which corresponds to 0.108 g of Co) was dissolved in 1 dm^3 of distilled water and 100 cm^3 of this solution were diluted to 1 dm^3 with distilled water. Volumes of 0.5, 1, 2 and 3 cm^3 of this solution were then taken and each treated as above from 'Then 0.5 cm^3 of concentrated HCl and six drops of concentrated HNO$_3$ were added and the solution.... ' The measured absorbances of these standard solutions were found to be 0.15, 0.23, 0.40 and 0.56 and the absorbance of the unknown found to be 0.20.

It was then possible to calculate the concentration of Co in the original sample in mg kg^{-1} (ppm) as follows.

Table 19 gives the absorbances for the standard solutions of cobalt in ppm. When A is plotted against [Co], a linear relationship is found, in keeping with the Beer–Lambert law, and a [Co] value of 0.15 ppm is found for the unknown.

Working back to the original 5-g sample, it is possible to calculate that there are [(0.15 × 50)/20] × 100 μg of Co that came from the original 5-g sample. This works out at 7.5 μg (g sample)$^{-1}$ or 7.5 mg (kg sample)$^{-1}$ or 7.5 ppm (on a w/w basis).

Table 19 Absorbances of the standard Co solutions

A	[Co](ppm)
0.15	0.11
0.23	0.22
0.40	0.44
0.56	0.66

It is instructive to compare this spectrophotometric method with that of flame atomic absorption spectrometry. The latter method is somewhat faster and possesses a comparable LOD. The analytical method involves taking 5 g of soil, ashing and taking up the residue in 66% HCl to a volume of 100 cm^3. Following an appropriate dilution, this can be aspirated into the flame at an operating wavelength of 240.7 nm. Calibration standards can be made up in the low ppm range with 66% HCl and again aspirated directly into the flame. The atomiser should be frequently flushed with distilled water since concentrated HCl is corrosive. In addition, a reagent blank should be run after every five samples.

3.10 GRAVIMETRIC DETERMINATION OF NICKEL IN STEEL

3.10.1 Summary

Nickel can be quantitatively determined in steel by a dissolution procedure followed by precipitation of the nickel–dimethylglyoxime chelate and its gravimetric determination. If carried out carefully by a trained analytical scientist, the method is accurate with a coefficient of variation of less than 1%, but it is time consuming.

3.10.2 Introduction

Gravimetric analysis is a highly accurate and precise method of determining the major and minor metallic constituents of samples such as ores (e.g. Ce in an ore, Section 3.3) and alloys. It can be used to establish the reliability of alternative instrumental methods such as atomic absorption spectrometry that have now largely replaced gravimetry for routine analyses. It involves the isolation of an analyte of known stoichiometry in a weighable form. The most common procedure is to precipitate the analyte from solution in the form of a sparingly soluble compound formed by the addition of a suitable reagent. The precipitate is filtered off, washed to remove coprecipitated impurities, dried and weighed. In some cases, ignition to an oxide provides a weighable form of more reliable stoichiometry.

3.10.3 Procedure

(a) Weigh accurately a 1-g sample of the nickel-containing steel into a 400-cm^3 beaker. Add 15 cm^3 of concentrated HCl and 2 cm^3 of concentrated HNO$_3$, cover with a watch-glass and heat to boiling. Maintain the temperature near boiling until the sample has dissolved and ignore slight residues of silica. Cool, dilute to about 50 cm^3 with distilled or deionized water and filter if necessary. Wash the filter-paper with small portions of hot water, adding the washings to the sample solution, and dilute to a final volume of 250 cm^3.

(b) Add 5 g of citric acid to the solution, neutralize with dilute aqueous ammonia and then make just acid (pH 4–5) with dilute HCl. Warm the solution to about 70°C and add 25 cm^3 of 1% dimethylglyoxime solution followed by dilute aqueous ammonia dropwise, with stirring, until the solution is slightly ammoniacal. Stir well and allow to stand on a steam-bath for 30 min and then at room temperature until cold for about 1 h.

(c) Filter the solution through a previously dried and weighed sintered-glass crucible. Test the filtrate for completeness of precipitation by adding a few drops of 1% dimethylgloxime solution. If more nickel–dimethylglyoxime complex forms, return the filtrate to the beaker, add 5 cm^3 more reagent and repeat the procedure commencing with digestion on the steam-bath. Finally, wash with cold distilled or deionised water until the washings are chloride-free. Dry at 120°C for 1 h, cool in a desiccator and weigh as Ni(C$_4$H$_7$O$_2$N$_2$)$_2$. The structure is shown as **VII** in Section 2.5.2. It is then possible to calculate the percentage of Ni in the sample. Ideally steel samples should be processed in parallel so that the overall mean, relative standard deviation and the confidence interval about the mean can be calculated.

Chapter 4

Selected Analytical Problems Involving Inorganic and Organometallic Analytes which Contain Elements from Groups IB–VIIB

4.1 DETERMINATION OF TRACE CONCENTRATIONS OF COPPER IN THE AQUEOUS ENVIRONMENT BY STRIPPING VOLTAMMETRY

4.1.1 Summary

Differential-pulse anodic stripping voltammetry can be used to determine directly nanomolar concentrations of copper in samples taken from the aqueous environment. The method is accurate when compared with atomic absorption spectrometry and possesses advantages such as rapidity and portability for on-site monitoring. The methodology discussed in Section 4.1 will yield information on the speciation of copper in the particular environmental sample, e.g. labile metal [$CuCO_3$, $Cu(H_2O)_4^{2+}$, etc], organically bound labile metal, lipid-soluble metal. A worked calculation on the determination of total copper in a sea-water sample is given.

4.1.2 Introduction

Many metals play an important part in terrestrial and aquatic ecosystems because they are essential or toxic to living organisms. Se, for example, possesses both these characteristics. Trace metal contamination can have disastrous consequences such as Minamata (Japan) with elevated levels of Hg in fish and Itai-Itai (Japan) with elevated levels of Cd in rice. In these cases metals were transported from industrial effluents to fishing areas and rice fields, respectively, with disastrous consequences such as suffering, disease and death of many people. Generally, concentrations of heavy metals in natural waters are very low, 10^{-11}–10^{-8} mol dm^{-3} (0.001–1 μg dm^{-3}) with the lowest levels obviously being detected in open seas.

Sampling and analytical operations demand careful and skilful handling to achieve reliable results. The risk of contamination of samples is present throughout the whole analytical method. The laboratory itself must be ultraclean with filtration of incoming air, clean benches, protective gloves of a non-dust collecting design, tacky mats at entrances, coverage of ceilings, walls and floor with plastic material and a minimum number of people in the laboratory with a minimum of movement. The vessels used for sampling and subsequent analytical operations should be made of quartz, PTFE, polyethylene or polypropylene. They should be soaked with acids and rinsed with ultrapure H$_2$O. The metal content of reagents should be checked and lowered, if necessary, by electrolysis, solvent extraction, etc. Laboratory staff should be dedicated towards obtaining reliable results. The overall analytical quality of a laboratory will be assessed by intercalibrations and by analyses of reference materials which are primarily of biological origin [e.g. NIST (Washington, DC, USA) Standard Reference Materials (SRMs)].

4.1.3 Procedure

The technique of differential-pulse anodic stripping voltammetry (DPASV) is used for Cu^{2+} determination and it has the facility to determine 10^{-11}–10^{-12} mol dm^{-3} concentrations of certain metal ions such as Cu^{2+} directly. A water sample is obtained, making sure that this sample has not been in contact with objects that might contaminate the sample during the gathering, e.g. ship-board sampling of surface waters. Such a sea-water sample is stored acidified to retard bacterial growth and adsorption of metals to surface of vessels. UV irradiation is initially used on the sample to oxidise interfering organic substances. These latter substances adsorb on the indicator electrode and inhibit electrochemical transfer at the electrode surface and can result in depressed peak currents, shifts in peak potentials and broader and anomalous peaks. Cu^{2+} is deposited at the hanging mercury drop electrode at a potential of ca -0.3 V in the form of Cu–Hg amalgam. Deposition is carried out for 10 min with a fixed stirring rate to ensure constant mass transport of Cu^{2+} to the electrode surface.

The concentration of metal deposited, $C_{Cu^{2+}}$, can be calculated from

$$C_{Cu^{2+}} = \frac{i_e t_e}{nFV} \qquad (26)$$

where i_e is the diffusion current during deposition time t_e, n is the number of electrons, F is the Faraday constant and V is the volume of the electrode.

Stripping of the deposited species is then carried out, after a quiescent period of 15–30 s, at a scan rate of 10 mV s^{-1} in the differential-pulse mode. The peak current of the metal stripped from the HMDE is then proportion to the metal concentration deposited, $C_{Cu^{2+}}$, which in turn is proportional to the metal concentration in bulk solution, $[Cu^{2+}]$. The unknown $[Cu^{2+}]$ is then calculated using the standard addition method of quantitation. Using this method of quantitation, the overall $[Cu^{2+}]$ in the sample is increased by ca 1 µg dm^{-3} for each standard addition.

The accuracy of this DPASV method can be checked against an AAS method. Flame AAS can be used following preconcentration of Cu^{2+} from the sea-water sample by a solvent extraction procedure involving extraction of a neutrally charged Cu–APDC (ammonium pyrrolidine dithiocarbamate) chelate from, say, 5 cm^3 of sea water into a small volume of methyl isobutyl ketone (e.g. 0.5 cm^3). Using 10-cm^3 quartz test-tubes with ground-quartz stoppers and 10-min shaking, complete extraction can be achieved. Such a liquid–liquid extraction is desirable in AAS since Cl^- from the sea-water matrix can interfere in the atomisation process. Although flame AAS can determine a larger number of elements, DPASV has fewer chemicals involved and hence less interferences, has less handling, has less time expenditure and is more portable for on-site monitoring.

An electroanalytical technique such as PDSAV can be extremely useful in yielding information of trace metal speciation in such samples from the aqueous environment, e.g. determination of $[Cu^{2+}]$ in its different physico-chemical forms in the sample such as particulate matter (with a diameter of > 450 nm retained by a 0.45-µm filter), dissolved matter [simple, relatively toxic, inorganic species $Cu(H_2O)_4^{2+}$ with a diameter of ca 1 nm], organic complexes (such as Cu–fulvic acid with a diameter of 2–4 nm) and relatively non-toxic Cu adsorbed on colloidal particles such as humic acid with a diameter of 10–500 nm. Lipid-soluble metal complexes are particularly toxic forms of heavy metals since they can diffuse through a biomembrane and carry both metal ion and ligand into the cell, e.g. Cu xanthates from mineral flotation plants and Cu–8-hydroxyquinolinate used as an agricultural fungicide. Labile ($CuCO_3$) or free Cu [e.g. $Cu(H_2O)_4^{2+}$] can, under certain conditions, correlate with that fraction of the total concentration in the water sample that is toxic to aquatic organisms and, as such, can be determined by the Cu that can be determined by DPASV. This percentage of total dissolved metal will be affected by electrochemical operating parameters such as deposition potential, stirring rate, Hg drop diameter, pH, temperature and supporting electrolyte. Under these conditions, labile metal consists of free metal ion and metal that can dissociate in the double layer from complexes or colloidal particles and hence be deposited on the HMDE. In the case of Cu^{2+}, Cu^{2+}–fulvic acid complexes are electrochemically non-labile whereas Cu^{2+} adsorbed on mixed inorganic organic colloids such as humic acid–Fe_2O_3 is of medium electrochemical lability.

Shifts in the ASV E_p values of metal ions in the presence of complexing agents can provide information about the thermodynamic stability of complexes in solution. Quantitative deductions for real samples from these shifts are impossible owing to the complexity of the sample with many unknown ligands and several metals. Some qualitative deductions can, however, be made, e.g. in ASV the E_p of Cu^{2+} in sea water is 0.2 V more negative than in a nitrate or acetate supporting electrolyte. This shift reflects the relatively high stability of Cu^+ chloro complexes compared with

Analytes with Group IB–VIIB Elements

Table 20 Electrochemical speciation scheme for Cu in water samples following filtration of an unacidified sample through a 0.45-μm membrane, rejection of particulates and investigation of the filtrate as indicated

Aliquot	Volume/cm^3	Operation	Interpretation
1	20	Acidify with 0.05 mol dm^{-3} HNO$_3$, add 0.1% H$_2$O$_2$ and UV irradiate for 8 h; ASV at pH 4.7	Total metal
2	10	ASV at natural pH for sea water. Add 0.025 mol dm^{-3} acetate buffer (pH 4.7) for fresh water	Labile metal
3	20	UV irradiation with 0.1% H$_2$O$_2$ at natural pH then ASV (not if [Fe] > 100 μg dm^{-3})	(3) − (2) ≡ organically bound labile metal
4	20	Pass through a small column of Chelex 100 resin. ASV of eluent	Very strongly bound metal
5	20	Extract with 5 cm^3 hexane–20% butan-1-ol. ASV of acidified, UV-irradiated aqueous phase (dissolved solvent in aqueous phase must be removed first)	(1) − (5) = lipid-soluble metal

those of Cu^{2+}. In sea water with high [Cl$^-$], Cu is stripped from the electrode in a 1e reaction to give Cu(I) chloro complexes, whereas in nitrate media, Cu^{2+} is produced in a 2e step.

This electroanalytical approach to speciation measurements has the limitation of being unable to measure the concentrations of individual ionic species, common to most speciation techniques such as solvent extraction and ion-exchange chromatography. ASV is a dynamic system that draws current through a solution and disturbs ionic equilibria. Direct electrochemical speciation procedures are limited to measuring gross behavioural differences, as illustrated in Table 20.

4.1.4 Calculation

A sample of sea water was divided to give two 20-cm^3 aliquots. Both of these were subjected to acidification with 0.05 mol dm^{-3} HNO$_3$, addition of 0.1% H$_2$O$_2$ and UV irradiation for 8 h. Both solutions were then made up to 25 cm^3 to give a pH of 4.7. ASV was then carried out on the first solution with deposition at -0.3 V for 10 min to give a stripping peak of 24.6 units in height, which corresponded to total copper in the sample. To the second solution was added 0.1 cm^3 of a standard solution of 5×10^{-6} mol dm^{-3} Cu^{2+}. An ASV peak of 39.8 units was found for this solution under identical operating conditions. Calculate the concentration of total copper in the sea-water sample in ppb given that the atomic mass of Cu is 63.55.

The standard addition equation that is appropriate to polarography and voltammetry is

$$C_u = \frac{i_1 v C_s}{(i_2 - i_1)V + i_2 v} \quad (27)$$

where C_u is the concentration of the unknown, C_s is the concentration of the standard, v is the volume of the standard addition, V is the volume of the sample in the voltammetric vessel, i_1 is the stripping peak of the unknown and i_2 is the stripping peak of the unknown plus standard addition. Therefore,

$$C_u = \frac{24.6 \times 0.1 \times 5 \times 10^{-6}}{(15.2 \times 25) + (39.8 \times 0.1)}$$

$$= \frac{12.3 \times 10^{-6}}{380 + 3.98}$$

$$= 3.20 \times 10^{-8} \text{ mol dm}^{-3}$$

Therefore, in the original sea-water sample, the concentration of total copper is

$$\frac{3.20 \times 25}{20} \times 10^{-8} \text{ mol dm}^{-3} = 63.55 \times 4.0 \times 10^{-8} \text{ g dm}^{-3}$$

$$= 2.54 \times 10^{-6} \text{ g dm}^{-3}$$

$$= 2.54 \text{ ppb}$$

Note that this method assumes linear calibration of peak height with concentration. This could be checked by making several such standard additions and plotting the graph shown in Figure 10. It also assumes that reagents such as HNO_3 and H_2O_2 contain no copper, hence a blank should be run to establish a baseline level for copper prior to its determination in sea water.

4.2 DETERMINATION OF ORGANOMERCURY COMPOUNDS IN FISH SAMPLES BY HPLC–COLD VAPOUR ATOMIC ABSORPTION SPECTROMETRY

4.2.1 Summary

The coupled technique high-performance liquid chromatography–cold vapour atomic absorption spectrometry can be used to separate and determine intact organomercury compounds such as $MeHg^+$ in fish samples. $MeHg^+$ can be determined in the concentration range 0–1.01 μg g^{-1} with an LOD of 0.6 ng. The technique was found to be superior to gas chromatography–atomic spectrometry since not only does the latter require derivatisation but there also exists the possibility of dealkylation/alkyl transfer reactions due to the presence of other organometallic species in the same sample.

4.2.2 Introduction

Mercury is used in a variety of products and industrial processes (Table 21) but in recent years its use has dwindled owing to concern over the effects of mercury on entering the environment. There have been several accidents, with fatalities from misuse or spillage of mercury compounds, in the last 20 years. In Minamata (Japan) the effluent of a chemical plant containing methylmercury entered the local bay and fish bioconcentrated the pollutant mercury compound.[55] Fatalities occurred after consumption of the contaminated fish. In Iraq (1971–72), consumption of seed sprayed with fungicidal methylmercury compounds resulted in 459 fatalities.

Problems associated with detecting methylmercury in fish or birds came to light in the late 1960s because the elevated concentrations found could not be attributed to

Analytes with Group IB–VIIB Elements

Table 21 Use of organomercury compounds

Compound[a]	Use	Comments[b]
CH_3HgX	Agricultural seed dressing, fungicide	Banned Sweden 1966, USA 1970, as seed disinfectant. Not used today in Europe or USA. Used in laboratories
C_2H_5HgX	Cereal seed treatment	Banned USA, Canada 1970. Used in UK
RHgX	Catalyst for urethane, vinyl acetate production	
C_6H_5HgX	Seed dressings, fungicide, slimicide, general bactericide. For pulp, paper, paints	Banned as slimicide USA 1970. Banned as rice seed dressing Japan 1970. Used in UK
p-$CH_3C_6H_5HgX$	Spermicide	
$ROCH_2CH_2HgX$	Seed dressings, fungicides	Banned Japan 1968. Used in UK
$ClCH_2CH(OCH_3)CH_2HgX$	Fungicide, pesticide, preservative	
Thiomersal	Antiseptic, C_2H_5Hg derivative	
Mercurochrome	Antiseptic, organomercury fluorescein derivative	
Mersalyl	Diuretic, methoxyalkyl derivative, $RCH_2CH(OCH_3)CH_2HgX$	Little used today. R = o-$COOHCH_2OC_6H_4CONH-$
Chlormerodrin	Diuretic, methoxyalkyl derivative, $NH_2CONHCH_2CH(OCH_3)CH_2HgCl$	Little used today
Mercarbolide	o-HOC_6H_4HgCl	o-Chloromercuriphenol
Mercurophen	o-NO_2-p-$ONaC_6H_3HgOH$	
Mercurophylline	Diuretic	

[a] X = anionic group; wide range of X known, e.g. OAc^-, PO_4^{3-}, Cl^-, $NHC(NH)NHCN^-$.
[b] UK status refers to usage up to early 1980s.

methylmercury spillage or misuse. It was later demonstrated that mercury can be methylated in the environment and accumulated in the tissue of fish or birds.[56] Upon entering the environment, mercury in any form and from a variety of sources (Table 21) can be transformed into its toxic methyl derivatives such as CH_3Hg^+ and $(CH_3)_2Hg$. The need to understand the biogeochemical cycling of mercury and the resultant speciation has led the search for analytical methods with suitable selectivities and low LODs. Analysis for total mercury in any matrix involves a digestion procedure, to transform the organometallic or otherwise bound mercury species into Hg^{2+}, followed by reduction by tin(II) or sodium tetrahydroborate and determination of the resultant elemental mercury by atomic spectrometry. Organic mercury is generally determined after extraction into organic solvents, stripping into aqueous solution for cleaning purposes, re-extraction into an organic solvent and chromatography.

Techniques for the detection and determination of organomercury compounds in environmental media now involve the coupling of chromatography (gas or liquid) with atomic spectrometry (atomic absorption, emission and fluorescence). Extensive cleaning procedures can hence be avoided and unambiguous identification and quantitation of organomercury compounds achieved. For the determination of organomercury compounds by GC–atomic spectrometry, derivatization steps and clean-up procedures are needed. The presence of other organometallic species in the same environmental samples may promote dealkylation or alkyl transfer reactions during these steps and/or during volatilization of the species in the gas chromatograph. Electron-capture detection can also be used with gas chromatography (absolute limits of detection are about 50–100 pg), but the use of extremely pure solvents is necessary, together with tedious cleaning procedures, to avoid co-elution of electron-capturing species with the organomercury compounds.[57] Inorganic mercury cannot be determined by gas chromatography. However, its concentration in the matrix under investigation can be derived from the difference between the total and organic mercury content. HPLC coupled with atomic spectrometry is found to effect the separation of organomercurials after simple extraction procedures, maintain the integrity of the original sample and provide adequate LODs. This is illustrated in the following procedure (Section 4.2.3).

4.2.3 Procedure[58]

A fine suspension of fish tissue was obtained by blending a known amount in water. This was then mixed with an equal amount of diatomaceous earth (Celite 545, acid-washed), transferred into a 22 mm i.d. chromatographic column and eluted with chloroform. The 20 cm^3 of chloroform eluent collected was extracted with 2 cm^3 of 0.01 mol dm^{-3} sodium thiosulphate. Methylmercury(II) thiosulphate was thus formed in the aqueous phase and did not need further treatment before determination by HPLC–cold vapour atomic absorption spectrometry.

HPLC was coupled with cold-vapour atomic-absorption spectrometry (as shown in Figure 26) for the separation and determination of selected organomercury compounds in fish.[57] A 25 cm long Zorbax ODS (5 μm) column was used with methanol–0.05 mol dm^{-3} ammonium acetate (60 + 40, v/v) containing 0.01% 2-mercaptoethanol. The eluate was burned in a copper tube and the elemental mercury produced was swept to the cold-vapour cell of the AAS instrument. The limit of detection, corresponding to a signal which was twice the standard deviation of the baseline noise, was quoted as 0.6 ng with linear calibration in the range 0–1.01 μg g^{-1} and a 93–106% recovery of spikes for the organomercury species MeHg$^+$.

4.3 DETERMINATION OF ZINC IN A PHARMACEUTICAL FORMULATION BY ION-EXCHANGE SEPARATION AND COMPLEXOMETRIC TITRATION

4.3.1 Summary

Zinc, as contained in a pharmaceutical mineral supplement, can be separated from other constituents of this complex matrix by its retention as the anionic chloro complex on an

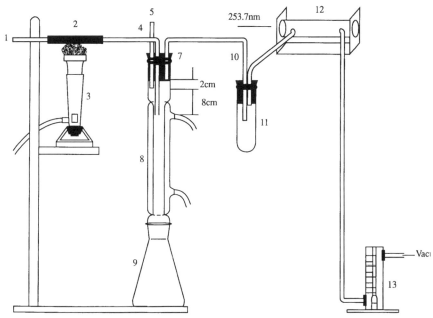

Figure 26 Atomic absorption detection of Hg vapour generated from the HPLC effluent. (1) HPLC effluent; (2) Cu tube; (3) Meker burner; (4), (5) and (6) narrow-bore HPLC tubing; (7) rubber stopper; (8) condenser; (9) Erlenmeyer flask; (10) narrow-bore PTFE tubing; (11) solvent trap; (12) atom cell; (13) flow meter. Reproduced by permission of the Royal Society of Chemistry from W. Holak, *Analyst*, 1982, **107**, 1457

anion-exchange resin prior to elution with HNO_3 and determination by EDTA using methylthymol blue indicator. The chemical structure of ion-exchange resins and the chemistry of the complexometric titration are discussed in the solution of this problem. The overall analytical method is simple and inexpensive to use but is subject to interference from other metals present in the sample that also form stable anionic chloro complexes and are titratable with EDTA. In this latter case, resort would have to be made to a more selective method of analysis using, for example, polarography or atomic absorption spectrometry.

4.3.2 Introduction

A widely used type of ion-exchange resin is a copolymer of styrene and divinylbenzene produced in bead form by suspension polymerisation in an aqueous medium:

The proportion of divinylbenzene is 2–20%, which results in a three-dimensional cross-linked structure that is rigid, porous and highly insoluble. A low degree of cross-linking produces beads which swell appreciably when in contact with a polar solvent and have large pores enabling ions to diffuse into the structure and exchange rapidly. Resins with a high degree of cross-linking have smaller pores and are more rigid. Swelling is less, the exchange process is slower and large ions may not be able to diffuse into the interior of the bead at all. Cation- or anion-exchanging properties are introduced into the resin by chemical modification after polymerisation, e.g. (i) the cation-exchanging group—SO_3^- (strong acid type) is introduced by reaction of the resin with chlorosulphonic acid resulting in mainly *para* substitution of the benzene rings; (ii) the anion-exchanging group —$\overset{+}{N}R_3$ (strong base type) is introduced by chloromethylating the resin followed by treatment with the appropriate amine, i.e.

[Reaction scheme: polystyrene + $ClCH_2OCH_3$ → chloromethylated polystyrene; then with R_3N → $CH_2-\overset{+}{N}R_3\ Cl^-$ substituted polystyrene]

When packed into a column, ion-exchange resin beads can selectively and quantitatively remove ions of a particular element or compound from a sample solution that is allowed to percolate slowly through the resin bed by a process of exchange of the counter ion [H_3O^+ in (i) and Cl^- in (ii)] for the particular ionic analyte. The retained species can be subsequently eluted by the passage of another solution through the column. Ion-exchange separations are used as preliminary stages in analytical procedures when one component of a sample interferes with the determination of another or to concentrate trace amounts of a species from dilute samples.

Many metals in solution can be determined by titration with a standard solution of a complexing/chelating agent such as EDTA (also known as H_4Y), used in the form of its readily soluble disodium salt, Na_2H_2Y. In the case of Zn^{2+},

$$Zn^{2+} + Y^{4-} = ZnY^{2-}$$

and the formation constant is

$$K_{ZnY^{2-}} = \frac{[ZnY^{2-}]}{[Zn^{2+}][Y^{4-}]} = 3.2 \times 10^{16}$$

This may be used to calculate the minimum pH required for quantitative reaction of Zn^{2+} with Y^{4-} since H_3O^+ ions are competing with Zn^{2+} for Y^{4-}.

$K'_{ZnY^{2-}}$ is defined as a conditional equilibrium constant in which allowance is made for the competing side-reaction:

$$K'_{ZnY^{2-}} = K_{ZnY^{2-}} \cdot \alpha_4$$

where

$$\alpha_4 = \frac{[Y^{4-}]}{C_L}$$

where C_L is the total amount of uncomplexed EDTA given by

$$C_L = [Y^{4-}] + [HY^{3-}] + [H_2Y^{2-}] + [H_3Y^-] + [H_4Y]$$

Therefore,

$$K'_{ZnY^{2-}} = \frac{[ZnY^{2-}]}{[Zn^{2+}]} \times \frac{1}{C_L}$$

$$= \frac{[ZnY^{2-}]}{[Zn^{2+}]C_L}$$

The variation of α_4 with pH is given in Table 22. Therefore, at pH 2:

$$K'_{ZnY^{2-}} = (3.2 \times 10^{16})(3.7 \times 10^{-14})$$
$$= 11.84 \times 10^2$$

at pH 3:

$$K'_{ZnY^{2-}} = (3.2 \times 10^{16})(2.5 \times 10^{-11})$$
$$= 8.0 \times 10^5$$

at pH 6:

$$K'_{ZnY^{2-}} = (3.2 \times 10^{16})(2.2 \times 10^{-5}) = 7.04 \times 10^{11}$$

Since a stoichiometric reaction requires $K'_{ZnY^{2-}} \geq 10^6$, a minimum pH in excess of 3 is required.

A metallochromic indicator such as methylthymol blue forms a coloured complex with Zn^{2+} but changes colour when Zn^{2+} is completely complexed by EDTA, as at the end-point in a titration.

Table 22 Variation of α_4 values with pH for EDTA

pH	α_4	pH	α_4
2	3.7×10^{-14}	7	4.8×10^{-4}
3	2.5×10^{-11}	8	5.4×10^{-3}
4	3.6×10^{-9}	9	5.2×10^{-2}
5	3.5×10^{-7}	10	3.5×10^{-1}
6	2.2×10^{-5}	11	8.5×10^{-1}
		12	9.8×10^{-1}

4.3.3 Procedure

(a) Prepare the anion-exchange resin by mixing about 10 g with 200 cm^3 of water in a 250-cm^3 beaker. Stir well and decant off most of the water to remove fine particles of resin. Transfer the resin to the column, making sure that the resin bed remains covered with water when it has settled. Pass 50 cm^3 of 1.5 mol dm^{-3} HCl through the column to condition it. *N.B.* During the operation of this procedure, do not allow the top of the resin bed to become dry.

(b) To a capsule/tablet (more than one may be required if the Zn content is less than about 2 mg) of the formulation in a 30-cm^3 beaker add 5 cm^3 of HCl–water (1 + 1). Cover the beaker with a watch-glass and heat it on a water-bath until decomposition is complete. If a residue remains, filter the hot solution through a moistened medium-porosity filter-paper using a Hirsch funnel to which suction can be applied. Collect the filtrate in or transfer it to a 25-cm^3 measuring cylinder. Rinse the beaker and filter-paper with several small portions of distilled or deionized water, adding the filtered washings to the sample solution until the volume in the measuring cylinder is 20 cm^3.

(c) Pass the capsule/tablet solution down the ion-exchange column at not more than 3 cm^3 min^{-1}, collecting the eluent in a 250-cm^3 conical flask. Rinse the measuring cylinder with five 10-cm^3 portions of 1.5 mol dm^{-3} HCl, passing each down the column and collecting the eluent in the flask. The zinc is retained on the column so the contents of the flask can be discarded.

(d) Pass 120 cm^3 of 0.5 mol dm^{-3} HNO$_3$, down the column at not more than 5 cm^3 min^{-1} to elute the anionic zinc chloro complex, collecting the eluent in a conical flask. Add 15 cm^3 of 4 mol dm^{-3} NaOH solution followed by about 2 g of hexamine. Adjust the pH of the solution to between 5.8 and 6.5 by adding more sodium hydroxide dropwise, then add enough methylthymol blue indicator to produce a distinct violet–blue coloration in the solution. Titrate the zinc with 0.01 mol dm^{-3} EDTA to a yellow end-point.

(e) Recondition the ion-exchange column by passing 100 cm^3 of 4 mol dm^{-3} HCl through it followed by 20 cm^3 of 1.5 mol dm^{-3} HCl. Other similar capsules/tablets can then be quantitatively analysed for their Zn content using procedures (b), (c), (d) and (e) in order to compile statistics on the particular batch, i.e. the mean content, relative standard deviation and the confidence interval about the mean.

4.4 VISIBLE SPECTROPHOTOMETRIC DETERMINATION OF BORON IN PLANT MATERIAL

4.4.1 Summary

Boron can be determined in plant material by visible spectrophotometry of its Azomethine H chelate at 430 nm following ashing, dissolution in 0.35 mol dm^{-3} H$_2$SO$_4$ and masking interfering metal ions. The method is particularly selective for the determination of boron in this and other complex matrices such as soils but is time consuming and requires the use of boron-free glassware. Reverse polarity capillary electrophoresis using a stacking technique can be used to determine the anionic chelate **XIV** at low ppb levels[61] in the presence of some 17 metals that do not interfere due to the presence of a masking buffer solution in the acidic run buffer.

4.4.2 Introduction

Boron usually occurs in small quantities in plants but variable amounts are quoted in the literature (e.g. 8–200 mg kg^{-1} for edible vegetables, quoted for a dried weight). The greatest amount usually occurs in the leaves. It is an essential nutrient for plants and deficiencies in beet, mangels, etc., are common. Sugar beet usually needs a heavy dressing of borax and if potatoes are grown shortly afterwards the potatoes may develop boron toxicity. Spinach is an indicator plant for deficiencies and some primulas are indicator plants for toxicity.

Particular effects attributed to boron are:

(i) it can delay the onset of calcium deficiency effects but cannot replace calcium;
(ii) it tends to keep calcium soluble;
(iii) it may act as a regulator of potassium/calcium ratios.

The availability of boron to the plant is decreased by liming and by dry conditions. Boron toxicity occurs in certain districts in California where soluble boron compounds are present in water used for irrigation.

A particular precaution necessary in assaying for boron is that all vessels, especially glassware, must be boron free. Pyrex is a borosilicate glass and must not be used. Boron can be determined spectrophotometrically using Azomethine H reagent [4-hydroxy-5-(salicylideneamino)-2,7-naphthalenedisulphonic acid]. This reagent forms a yellow chelate with borates in aqueous media[59,60]:

(XIV)

4.4.3 Procedure

Weigh 1.2 g of dried plant material into a porcelain crucible and add 50 mg of $Ca(OH)_2$. Ash at 500 °C for 4 h. Cool and remove from the furnace. Take up in 0.35 mol dm^{-3} H_2SO_4, filter through Whatman No. 2 filter-paper and dilute to 50 cm^3 with 0.35 mol dm^{-3} H_2SO_4. Pipette 1 cm^3 of the ash solution into a test-tube and add 2 cm^3 of the buffer masking solution (250 g of ammonium acetate, 25 g of ethylene dinitrilotetraacetic acid, 10 g of the disodium salt of nitrilotriacetic acid dissolved in 400 cm^3 of H_2O with 125 cm^3 of CH_3COOH added slowly to this with thorough mixing). Next add 1 cm^3 of Azomethine H reagent solution (0.9 g of Azomethine H reagent and 2 g of ascorbic acid dissolved in 50 cm^3 of H_2O. Heat gently and make up to 100 cm^3 with H_2O. This must be prepared freshly just prior to use). Mix well and allow to stand for 2 h for full colour development. Compare the absorbance of the unknown with standards using a spectrophotometer, wavelength 430 nm, with 1-cm optical cells. Boron standards of 0–4 ppm B are prepared in 0.35 mol dm^{-3} H_2SO_4 from a stock solution of 1000 ppm B as boric acid in H_2O (5.7134 g of H_3BO_3 in 1 dm^3 of distilled water). These standards are subjected to colour development with Azomethine H reagent as above. Results are quoted as ppm B in dry plant material.

Recent studies by Oxspring et al.[61] on reverse polarity capillary electrophoresis of the anionic chelate **XIV** using a stacking technique have lowered the LOD of the visible spectrophotometric method to low ppb concentrations. Linear calibration is observed with a correlation coefficient, r, of 0.9909. Again, the technique is particularly selective for boron in an artificial water matrix that contains some 17 metals at the ppb concentrations that can be encountered in river waters. In this case, the masking buffer solution is added to the CE run buffer and detection is with a diode-array detector. An electropherogram of 25 ppb B using a detection wavelength of 410 nm is given in Figure 27.

4.4.4 Calculation

A 7.94 g sample of a plant was dry ashed and the residue was dissolved in 0.35 mol dm^{-3} H_2SO_4 and made up to 250 cm^3 with 0.35 mol dm^{-3} H_2SO_4 to give solution A. A 50-cm^3 volume of this solution was diluted with 20 cm^3 of a stock Azomethine H solution and 30 cm^3 of buffer masking agent and allowed to stand for 2 h for full colour development of the Azomethine H–boron chelate. The resulting yellow solution gave an absorbance maximum at 430 nm and an absorbance value of 0.364 in a 1-cm cell. A further 50 cm^3 of solution A were diluted with 20 cm^3 of the stock Azomethine H solution, 26 cm^3 of buffer masking agent and 4 cm^3 of a 3 ppm standard boron solution. This solution was also allowed to stand for 2 h and then gave an absorbance of 0.688 at 430 nm in a 1-cm cell. Calculate the percentage of B in the original plant sample and also express the result in ppm.

Using the Beer–Lambert law for the two solutions subjected to visible spectrophotometry,

$$0.364 = \varepsilon \left(C_u \times \frac{50}{100} \right) \times 1 \qquad (28)$$

Analytes with Group IB–VIIB Elements

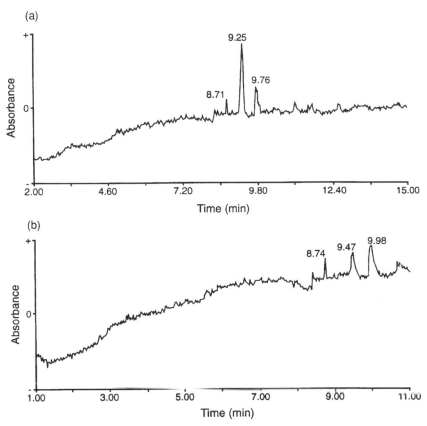

Figure 27 Reverse polarity capillary electrophoresis using a stacking technique of the anionic chelate of boron with Azomethine H in an artificial river water matrix. (a) 25 ppb boron with chelate peak observed at 9.25 min; (b) artificial river water matrix without boron with peak at 9.47 min due to boron leached from glassware. Reproduced by permission of Elsevier Science from D. A. Oxspring *et al.*, *Anal. Chim. Acta*, in press

$$0.688 = \varepsilon\left(C_u \times \frac{50}{100} + 3 \times \frac{4}{100}\right) \times 1 \qquad (29)$$

Taking equation (29) and solving for C_u, the concentration of boron in solution A

$$0.688 = \frac{0.364}{C_u \times \frac{50}{100}}\left(C_u \times \frac{50}{100} + 3 \times \frac{4}{100}\right)$$

$$0.688 = 0.364 + \frac{0.364}{C_u \times \frac{50}{100}}\left(\frac{12}{100}\right)$$

$$C_u = \frac{0.364 \times 12 \times 100}{50 \times 100 \times (0.688 - 0.364)} = 0.27 \text{ ppm}$$

which is a boron concentration of 0.27 mg dm^{-3}.

In 50 cm^3 solution there are $\frac{0.27}{2 \times 10}$ mg of boron

In 250 cm^3 solution there are $\frac{0.27}{2 \times 10} \times 5$ mg of boron

Therefore, i.e. $\frac{0.27}{2 \times 10} \times 5 \times 10^{-3}$ g

$$\text{B percentage in sample} = \frac{0.27}{2 \times 10} \times \frac{5 \times 10^{-3}}{7.94} \times 100\%$$
$$= 8.5 \times 10^{-4}\%$$
$$= 8.5 \times 10^{-6} \text{ g of B in 1 g of sample}$$
$$= 8.5 \text{ ppm B}$$

4.5 DETERMINATION OF ORGANOLEAD COMPOUNDS IN AIR SAMPLES BY GAS CHROMATOGRAPHY–ATOMIC ABSORPTION SPECTROMETRY

4.5.1 Summary

The coupled technique gas chromatography–atomic absorption spectrometry can be used to monitor the complete speciation of alkyllead compounds in air samples. R_4Pb species can be trapped on a solid adsorbent prior to thermal desorption into the GC–AAS system and R_3Pb^+ and R_2Pb^{2+} species are collected in water, solvent extracted into hexane or benzene, followed by alkylation and determination by GC–AAS. LODs were estimated to be in the range 0.01 ng of Pb for Me_4Pb to 0.08 ng of Pb for Et_2Pr_2Pb. Concentrations of 1.1 ng m^{-3} Me_4Pb and 0.5 ng m^{-3} Me_3Pb^+ have been determined in an air sample by this procedure.[64]

4.5.2 Introduction

Environmental pollution by organic lead is almost entirely due to the manufacture and use of toxic tetraalkyl lead (R_4Pb) compounds as petrol (gasoline) additives in order to increase the octane rating of fuels for high-compression internal combustion engines. In the environment R_4Pb compounds decompose to inorganic Pb with R_3Pb^+ and R_2Pb^{2+} compounds as fairly persistent intermediates. Concern over lead pollution has resulted in ever more stringent limits on the organolead content of petrol in many countries and the availability of lead-free petrol. Species-sensitive analytical techniques are therefore required for the determination of organolead compounds in environment samples. Many of the methods capable of measuring organic lead are either not specific or not sensitive enough for environmental application, are generally cumbersome and may involve several analysis steps in order to yield complete information in the various species.[62] The volatility of alkyllead compound permits their separation by chromatography but the use of conventional detectors such as flame ionisation is precluded by the interference from hydrocarbons present at higher concentrations. Interfacing chromatography with element-specific atomic absorption spectrometry provides one of the most specific and low

LOD analytical techniques for organolead compounds. GC–AAS has been the most widely used chromatography–atomic spectrometry system for the specific determination of alkyllead compounds. Glass and stainless-steel columns have been used and the transfer lines are generally heated. GC ovens have been operated in isothermal and temperature-programmed mode. The greatest sensitivity has been obtained with systems employing a silica furnace detector (SFD) which has been either electrothermally heated or suspended in an air–acetylene flame above the burner head and in the path of the light beam.

4.5.3 Procedure

Gas-phase organolead compounds are separated from particulate matter by filtration, followed by trapping the R_4Pb species from the filtered air stream with a solid adsorbent held at low temperature and then thermal desorption into the GC–AAS system for analysis. Gaseous R_3Pb^+ and R_2Pb^{2+} compounds are collected in water and aqueous samples are extracted with an organic solvent (hexane or benzene). These extracts may be alkylated and analysed by injecting aliquots into the GC–AAS system. Organolead compounds in the atmosphere may be present in particulate matter and in the vapour phase. Particle associated R_4Pb can be leached from the filter into an organic solvent followed by analysis by GC–AAS.

The GC–AAS system shown in Figure 28 has been employed for the complete speciation of alkyllead in air samples.[63] The reproducibility of the system was studied by injecting seven aliquots of a standard containing 1 ng of lead for each R_4Pb compound and recording the peak heights. The relative standard deviations were between 7 and 12%. Calibration data were obtained by measurement of a series of standards containing alkyllead between 0.1 and 1 ng (as Pb). A linear response was found in all cases. A typical GC–AAS trace of R_4Pb and ionic propylated alkyl lead compounds is shown in Figure 29. Detection limits were estimated to be in the range from 0.01 ng of Pb for Me_4Pb to 0.08 ng of Pb for Et_2Pr_2Pb for peak-height measurements and definition of the detection limit as three times the standard deviation of the baseline noise, divided by the sensitivity. For the five R_4Pb compounds (R = Me, Et or their combinations) it was found that equal quantities of Pb gave equal peak areas ($\pm 4\%$) and peak areas were employed when analysing environmental samples. Concentrations of 1.1 ng m^{-3} Me_4Pb and 0.5 ng m^{-3} Me_3Pb^+ have been determined in an air sample by this procedure.[64]

4.6 DETERMINATION OF NITROGEN DIOXIDE IN AIR SAMPLES BY SORBENT TUBE COLLECTION–COLORIMETRY

4.6.1 Summary

Air is sampled using a sampling pump and a sorbent tube containing molecular sieve material which traps NO_2. Following desorption, NO_2 is determined by reaction with

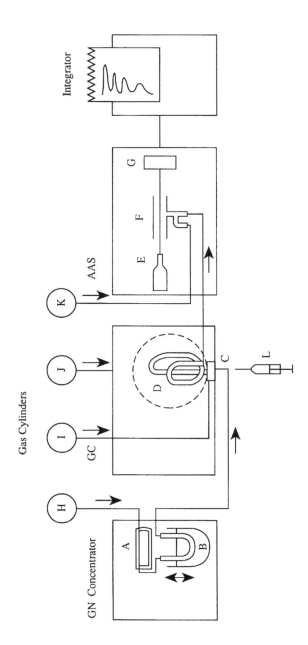

Figure 28 GC–AAS instrumentation. (A) adsorption tube in oven; (B) cryogenic trap; (C) injection port; (D) GC column inside oven; (E) Pb hollow-cathode lamp; (F) electrothermally heated silica tube; (G) monochromator; (H) He; (I) N_2; (J) air; (K) H_2; (L) manual injection. Reproduced with permission.

Analytes with Group IB–VIIB Elements

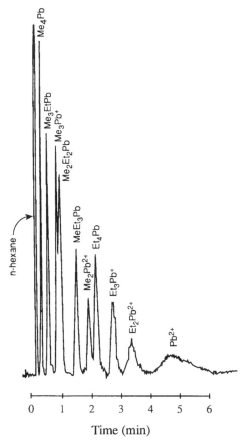

Figure 29 GC–AAS trace of R_4Pb and propylated R_3Pb^+ and R_2Pb^{2+} compounds without background correction. Reproduced with permission

sulphanilic acid to yield a diazo product followed by its reaction with *N*-(1-naphthyl)ethylenediamine to yield a coloured azo dye which absorbs visible radiation at 550 nm. The concentration of NO_2 can be expressed as ppm, a measure of concentration by volume, or as mg m^{-3}, a measure of concentration by mass. These concentrations can then be compared with published occupational exposure limits in order to assess compliance with The Control of Substances Hazardous to Health (COSHH) Regulations (1994) in the UK.

4.6.2 Introduction

Exposure of employees to substances hazardous to health should be prevented or, where this is not reasonably practicable, adequately controlled. This is a fundamental requirement of the COSHH Regulations (1994) in the UK. Exposure can occur by inhalation, ingestion or absorption through the skin, but inhalation is usually the main route of entry into the body. The Health and Safety Executive publication EH 40/95 lists the occupational exposure limits which should be used in determining the adequacy of control of exposure by inhalation, as required by the COSHH regulations.

There are two types of occupational exposure limit, the maximum exposure limit (MEL) and the occupational exposure standard (OES). An OES is set at a level at which there is no indication of risk to health. A listing is given in Table 2 of the EH 40/95 publication (1995 edition, p. 29). A substance is assigned an MEL if there are serious implications for the health of a small population of workers or there is the risk of relatively minor health effects in a large population. In practice, MELs have most often been allocated to carcinogens and to other substances for which no threshold of effect can be identified and for which there is no doubt about the seriousness of the effects of the exposure. Examples include acrylonitrile, carbon disulphide, ethylene oxide, total isocyanates, nickel and vinyl chloride. These are listed in Table 1 of the EH 40/95 publication (1995 edition, p. 27).

The pattern of effects due to exposure to substances hazardous to health varies considerably depending on the nature of the substances and the exposure. Some effects require prolonged or accumulated exposure. The long-term [8-h time-weighted average (TWA)] exposure limit is intended to control such effects by restricting the total intake by inhalation over one or more workshifts. Other effects may be seen after brief exposures. Short-term exposure limits (usually 15 min) may be applied to control these effects. Where long term limits also apply, the short-term limits restrict the magnitude of excursions above the average concentration during longer exposures. Examples of occupational exposure standards (OESs) for both long and short-term exposure are given in Table 23, including nitrogen dioxide as an example. Table 24 gives examples of MELs for both long- and short-term exposure and includes isocyanates, which are the subject of Section 6.1.

The long-term exposure of 8-h TWA may be represented mathematically as

$$8\text{-h TWA} = \frac{C_1 T_1 + C_2 T_2 + \cdots + C_n T_n}{8} \tag{30}$$

where C_1 is the occupational exposure and T_1 is the associated exposure time in hours in any 24-h period.

For example, if a worker spent 2.5 h at an exposure of 0.12 mg m^{-3} to a particular substance, then 2 h at 0.07 mg m^{-3}, a further 2 h at 0.20 mg m^{-3} and a final 1.5 h of the working 8-h day at 0.10 mg m^{-3}, the 8-h TWA would be

$$\frac{2.5(0.12) + 2(0.07) + 2(0.20) + 1.5(0.10)}{8} = 0.12 \text{ mg m}^{-3}$$

Long- and short-term exposures to gases and vapours are expressed in ppm concentrations (a measure of concentration by volume) and mg m^{-3} concentrations (a measure of concentration by mass). In converting from ppm to mg m^{-3}, a tempe-

Table 23 Examples of occupational exposure standards (OESs). Crown copyright. Reproduced with the permission of the Controller of HMSO from EH 40/95, *Occupational Exposure Limits 1995*

Substance	Formula	CAS No.	Long-term exposure limit (8-h TWA reference period)		Short-term exposure limit (15-min reference period)	
			ppm	mg m^{-3}	ppm	mg m^{-3}
N-Methyl-N,2,4,6-tetranitroaniline	$(NO_2)_3C_6H_2N(NO_2)CH_3$	479-45-8	—	1.5	—	3
Mevinphos (ISO)	$C_7H_{13}O_6P$	7786-34-7	0.01	0.1	0.03	0.3
Mica		12001-26-2				
Total inhalable dust			—	10	—	—
Respirable dust			—	1	—	—
Molybdenum compounds (as Mo)	Mo					
Soluble compounds			—	5	—	10
Insoluble compounds			—	10	—	20
Monochloroacetic acid	$ClCH_2CO_2H$	79-11-8	0.3	1	—	—
Morpholine	C_4H_9NO	110-91-8	20	70	30	105
Naled (ISO)	$C_4H_7Br_2Cl_2O_4P$	300-76-5	—	3	—	6
Naphthalene	$C_{10}H_8$	91-20-3	10	50	15	75
Nickel, organic compounds (as Ni)	Ni	7440-02-0	—	1	—	3
Nicotine	$C_{10}H_{14}N_2$	54-11-5	—	0.5	—	1.5
Nitric acid	HNO_3	7697-37-2	2	5	4	10
4-Nitroaniline	$NO_2C_6H_4NH_2$	100-01-6	—	6	—	—
Nitrobenzene	$C_6H_5NO_2$	98-95-3	1	5	2	10
Nitroethane	$C_2H_5NO_2$	79-24-3	100	310	—	—
Nitrogen dioxide	NO_2	10102-44-0	3	5	5	9
Nitrogen monoxide	NO	10102-43-9	25	30	35	45
Nitrogen trifluoride	NF_3	7783-54-2	10	30	15	45
Nitromethane	CH_3NO_2	75-52-5	100	250	150	375
1-Nitropropane	$C_3H_7NO_2$	108-03-2	25	90	—	—
Nitrotoluene, all isomers	$CH_3C_6H_4NO_2$		5	30	10	60

(continued overleaf)

Table 23. (continued)

Substance	Formula	CAS No.	Long-term exposure limit (8-h TWA reference period)		Short-term exposure limit (15-min reference period)	
			ppm	mg m^{-3}	ppm	mg m^{-3}
Octachloronaphthalene	$C_{10}Cl_8$	2234-13-1	—	0.1	—	0.3
n-Octane	$CH_3(CH_2)_6CH_3$	111-65-9	300	1450	375	1800
Oil mist, mineral			—	5	—	10
Orthophosphoric acid	H_3PO_4	7664-38-2	—	—	—	2
Osmium tetraoxide (as Os)	OsO_4	20816-12-0	0.0002	0.002	0.0006	0.006
Oxalic acid	COOHCOOH	144-62-7	—	1	—	2
Oxalonitrile	$(CN)_2$	460-19-5	10	20	—	—
2,2′-Oxydiethanol	$(HOCH_2CH_2)_2O$	111-46-6	23	100	—	—
Ozone	O_3	10028-15-6	0.1	0.2	0.3	0.6
Paracetamol	$C_8H_9NO_2$	103-90-2				
Total inhalable dust			—	10	—	—
Paraffin wax, fume		8002-74-2	—	2	—	6
Paraquat dichloride (ISO)	$[CH_3(C_5H_4N^+)_2CH_3][Cl^-]_2$	1910-42-5				
Respirable dust			—	0.1	—	—
Parathion (ISO)	$(C_2H_5O)_2PSOC_6H_4NO_2$	56-38-2	—	0.1	—	0.3
Parathion-methyl (ISO)	$C_8H_{10}NO_5PS$	298-00-0	—	0.2	—	0.6

Name	Formula	CAS			
Pentacarbonyliron (as Fe)	Fe(CO)$_5$	13463-40-6	0.01	0.08	—
Pentachlorophenol	C$_6$Cl$_5$OH	87-86-5	—	0.5	—
Pentaerythritol	C(CH$_2$OH)$_4$	115-77-5			1.5
Total inhalable dust			—	10	—
Respirable dust			—	5	20
Pentane, all isomers	C$_5$H$_{12}$	109-66-0	600	1800	750	2250
Pentan-2-one	CH$_3$COC$_3$H$_7$	107-87-9	200	700	250	875
Pentan-3-one	C$_2$H$_5$COC$_2$H$_5$	96-22-0	200	700	250	875
Pentyl acetate	CH$_3$COOC$_5$H$_{11}$	628-63-7	100	530	150	800
Perchloryl fluoride	ClO$_3$F	7616-94-6	3	14	6	28
Phenol	C$_6$H$_5$OH	108-95-2	5	19	10	38
p-Phenylenediamine	C$_6$H$_4$(NH$_2$)$_2$	106-50-3	—	0.1	—	—
Phenyl 2,3-epoxypropyl ether	C$_6$H$_5$OCH$_2$CHCH$_2$\O/	122-60-1	1	6	—	—
2-Phenylpropene	C$_6$H$_5$C(CH$_3$)=CH$_2$	98-83-9	—	—	100	480
Phorate (ISO)	C$_7$H$_{17}$O$_2$PS$_3$	298-02-2	—	0.05	—	0.2
Phosgene	COCl$_2$	75-44-5	0.02	0.08	0.06	0.25
Phosphine	PH$_3$	7803-51-2	—	—	0.3	0.4
Phosphorus, yellow	P$_4$	7723-14-0	—	0.1	—	0.3

Table 24 Examples of maximum exposure limits (MELs). Crown copyright. Reproduced with the permission of the Controller of HMSO from EH 40/95, *Occupational Exposure Limits 1995*

Substance	Formula	CAS No.	Long-term exposure limit (8-h TWA reference period)		Short-term exposure limit (15-min reference period)	
			ppm	mg m^{-3}	ppm	mg m^{-3}
2-Ethoxyethanol	$C_2H_5OCH_2CH_2OH$	110-80-5	10	37	—	—
2-Ethoxyethyl acetate	$C_2H_5OCH_2CH_2OOCCH_3$	111-15-9	10	54	—	—
Ethylene oxide	CH_2CH_2O	75-21-8	5	10	—	—
Formaldehyde	HCHO	50-00-0	2	2.5	2	2.5
Grain dust			—	10	—	—
Hydrogen cyanide	HCN	74-90-8	—	—	10	10
Isocyanates, all (as NCO)			—	0.02	—	0.07
Man-made mineral fibre[a]				5		
2-Methoxyethanol	$CH_3OCH_2CH_2OH$	109-86-4	5	16	—	—
2-Methoxyethyl acetate	$CH_3COOCH_2CH_2OCH_3$	110-49-6	5	24	—	—
4,4′-Methylenedianiline	$C_{13}H_{14}N_2$	101-77-9	0.01	0.08	—	—

Substance	Formula	CAS	ppm (TWA)	mg/m³ (TWA)	ppm (STEL)	mg/m³ (STEL)
Nickel	Ni		—	0.5	—	—
Nickel, inorganic compounds (as Ni)	Ni					
Soluble compounds			—	0.1	—	—
Insoluble compounds			—	0.5	—	—
2-Nitropropane	$CH_3CH(NO_2)CH_3$	79-46-9	5	18	—	—
Rubber fume[b]			—	0.6	—	—
Rubber process dust			—	6	—	—
Silica, crystalline	SiO_2					
Respirable dust			—	0.4	—	—
Styrene	$C_6H_5CH=CH_2$	100-42-5	100	420	250	1050
1,1,1-Trichloroethane	CH_3CCl_3	71-55-6	350	1900	450	2450
Trichloroethylene	$CCl_2=CHCl$	79-01-6	100	535	150	802
Vinyl chloride[c]	$CH_2=CHCl$	75-01-4	7	—	—	—
Vinylidene chloride	$CH_2=CCl_2$	75-35-4	10	40	—	—
Wood dust (hard wood)			—	5	—	—

[a] In addition to the maximum exposure limit specified above, man-made mineral fibre is also subject to a maximum exposure limit of 2 fibres ml^{-1}, 8-h TWA, when measured by a method approved by the Health and Safety Commission.
[b] Limit relates to cyclohexane-soluble material.
[c] Vinyl chloride is also subject to an overriding annual maximum exposure limit of 3 ppm.

rature of 25 °C and an atmospheric pressure of 1 bar are used.

For example, the 15-min short-term exposure limit for NO_2 is 5 ppm. What is this concentration expressed as mg m^{-3}?

Using the ideal gas equation:

$$PV = nRT \tag{31}$$

Rearranging equation (31) and multiplying both sides by molecular weight M:

$$\frac{nM}{V} = \frac{PM}{RT}$$

The left-hand side of this equation represents

$$\frac{moles}{dm^3} \times \frac{grams}{mole} = \frac{grams}{dm^3} = density$$

Therefore,

$$density = \frac{PM}{RT}$$

and

$$density\ of\ NO_2\ in\ g\ dm^{-3} = \frac{1 \times 46}{0.0821 \times 298}$$
$$= 1.88$$

5 ppm NO_2 is 5 cm^3 NO_2 per 10^6 cm^3 air = 9.4×10^{-3} g NO_2 per 10^6 cm^3 air

$$= \frac{9.4 \times 10^{-3}}{10^6} g\ NO_2\ per\ cm^3\ air$$

$$= 9.4 \times 10^{-3}\ g\ NO_2\ per\ m^3\ air$$

$$= 9.4\ mg\ NO_2\ m^{-3}\ (or\ 9\ mg\ NO_2\ m^{-3}\ to$$

the nearest whole mg)

The Health and Safety Executive in the UK also publishes recommended analytical methods for the determination of these hazardous airborne substances. In the USA, the National Institute of Occupational Safety and Health (NIOSH) and the Occupational Safety and Health Administration (OSHA) publish limits for airborne contaminants and the corresponding analytical methods for workplace exposure.

NO_2 is a major air pollutant because it is a product of all combustion processes (fireplace, cigarette smoke, petrol and diesel engine exhaust, power plants, industrial boilers). Smog is produced when hydrocarbons and NO_2 containing air are exposed to light: peroxyacetyl nitrate is formed in this Los Angeles-type smog. In addition, $NO_2 + H_2O \rightarrow HNO_3$, which is a component of acid rain. NO_2 is toxic to plants and animals. Sometimes it can be seen as a layer of brown air trapped along a city street. Its determination in air samples is therefore of significance.

4.6.3 Procedure

The sorbent tube is used to sample the particular air sample. In the case of NO_2, a glass tube containing two layers of solid adsorbent material (molecular sieve material) is used. When the air sample is actively pulled through this tube, airborne NO_2 is trapped in the first adsorbent layer with the second layer acting as a backup and indicator of sample breakthrough. After sampling, NO_2 is desorbed prior to its determination by colorimetry, in which NO_2 first reacts with sulphanilic acid (**XV**) to yield a diazonium salt (**XVI**), which then reacts with N-(1-naphthyl)ethylenediamine to give a azo dye (**XVII**), which can be determined by visible spectrophotometry at 550 nm.

NO_2 can also be determined by a badge method (e.g. ChemSense electronic reader for Noxosense passive dosimeter) or by a colour detector tube (e.g. Gastec detector tube).

4.6.4 Calculation

An air sample from a living site close to a power station generating NO_2 was obtained by pumping the air at a rate of 5 dm^3 min^{-1} for 30 min into 200 cm^3 of a reaction mixture containing **XV** and N-(1-naphthyl)ethylenediamine. The absorbance of the resulting solution was measured at 550 nm after the reaction reached completion and a value of $A = 0.60$ was obtained. Calculate [NO_2] in the polluted air in ppm (a measure of concentration by volume) and mg m^{-3} (a measure of concentration by mass), given that calibration is performed by bubbling certain volumes of NO_2 gas at NTP into 200 cm^3 of aqueous CH_3COOH containing reactants sulphanilic acid (**XV**) and N-(1-naphthyl)ethylenediamine. The reaction is allowed to reach completion in each case and the azo dye product (**XVII**) is monitored at 550 nm using visible spectrophotometry. The calibration data in Table 25 are obtained.

Also calculate the LOD of the procedure when the LOD is defined as that concentration of NO_2 which gives a signal twice that of the blank.

A calibration plot of cm^3 NO_2 vs A is plotted to give 1.8 cm^3 NO_2 for the unknown solution. Therefore,

1.8 cm^3 of NO_2 is contained in 150 dm^3 of air;

1.8 cm^3 of NO_2 is contained in 150×10^3 cm^3 of air;

12 cm^3 of NO_2 is contained in 10^6 cm^3 of air;

i.e. 12 ppm NO_2.

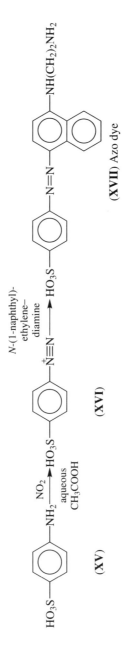

Table 25 Variation of absorbance of the azo dye **XVII** with calibration volumes of NO_2 initially taken for reaction to produce **XVII**

NO_2 (cm^3)	A
0	0.10
0.35	0.20
0.80	0.30
1.40	0.50
2.10	0.70

From the calculation given in Section 4.6.2:

$$12 \text{ ppm} = 22.56 \text{ mg NO}_2 \text{ m}^{-3}$$

It should be noted that this concentration exceeds the occupational exposure standards, Health and Safety Executive (UK) 1995 of 5 mg m^{-3} as an 8-h TWA or 9 mg m^{-3} as a 15-min short-term exposure limit.

The LOD will be that NO_2 concentration which corresponds to $A = 0.2$, i.e. 0.35 cm^3 NO_2, which is equivalent to 2.3 ppm NO_2.

4.7 DETERMINATION OF ARSENIC IN HAIR SAMPLES BY NEUTRON ACTIVATION ANALYSIS AND IN WALLPAPER BY X-RAY FLUORESCENCE SPECTROMETRY

4.7.1 Summary

Neutron activation analysis has been applied to measure total As levels in a sample of Napoleon's hair and resulted in the conclusion that 1.4 ± 0.2 ppm As does not represent a particularly high level when control values in the range 0.079–0.67 ppm are taken into account. Moreover, the As loading of Napoleon's wallpaper, as determined by X-ray fluorescence spectrometry, was not found excessive but could have contributed to his illness on St Helena, since on damp wallpaper many moulds can metabolise As compounds to the volatile and toxic $As(CH_3)_3$. These results would suggest that Napoleon was not deliberately poisoned.

4.7.2 Introduction

The element As is found in nature in minerals such as As_2S_3 and in organic compounds such as trimethylarsine [$(CH_3)_3As$]. Arsenic compounds are also used in agriculture and industry as herbicides, pesticides and wood preservatives with about 70% entering the environment as inorganic salts such as sodium arsenite (Na_3AsO_3) and lead arsenate [$Pb_3(AsO_4)_2$] and about 30% entering as organoarsonicals such as monomethylarsonic acid [$CH_3As(=O)(OH)_2$] and dimethylarsinic acid [$(CH_3)_2$

As(=O)OH]. Compounds such as phenylarsonic acid [PhAs(=O)(OH)$_2$] are used as veterinary antibiotics. Each of these inorganic and organic arsenic compounds differs with respect to toxicity and there can be interconversion of the compounds by biological and chemical action. As a result, it is becoming increasingly important to be able to monitor the chemical form in which the As is present in an environmental sample as well as the total As concentration. Total As concentrations of 0.02–7.5 mg dm^{-3} are toxic to plants, 5–50 mg day^{-1} are toxic and 100–300 mg day^{-1} are lethal to man.[2] Total As concentrations estimated in complex matrices are 1.7×10^{-3}–0.09 mg kg^{-1} (human blood), 6 mg kg^{-1} (soil), 3.7×10^{-3} mg kg^{-1} (sea water), 1.5–53 ng m^{-3} (air), 0.01–1.5 mg kg^{-1} (edible vegetables as per dried sample) and 0.2–10 mg kg^{-1} (marine fish as per dried sample).[2] It can be concentrated by marine fish (in Fiji, sea dwellers are supposed to produce more boys than girls in comparison with inland vegetarians, supposedly owing to the non-poisonous form of As in shellfish) and can accumulate in mammalian hair and nails (the case of Napoleon's hair and its analysis is pertinent).

A scheme to determine inorganic and organic As compounds in an environmental sample, which may contain particulates, volatiles and solution phase species, is given in Figure 30. ASS can be conveniently used to determine total As concentrations. The flame method has an LOD of 1000 ppb (or 250 ppb with arsine generation), which can be lowered to 50 ppb using concentration of As in a sample by solvent extraction of the ammonium pyrrolidine dithiocarbamate chelate into a relatively small volume of methyl isobutyl ketone.[39] The carbon rod atomiser has an LOD of 20 ppb (for 5-µl samples) which has to be used for As determination in drinking water,[39] for which the EC has set a mandatory limit of 50 ppb after simple physical treatment (filtration) and disinfection.

Total As concentrations can also be determined by neutron activation analysis and X-ray fluorescence analysis and these two techniques have been utilised in a very interesting example of forensic analysis, i.e. that performed on samples of Napoleon's hair and his wallpaper with the resulting conjectures as to the reasons for his death on St Helena in May 1821. An autopsy by his physician, Autommarchi, in the presence of Sir Thomas Reade, some staff officers and eight medical men revealed a cancerous growth of his stomach.[73] His symptoms have been compared with those of arsenic poisoning,[74-76] and indeed As was used as a rat poison in the early part of the nineteenth century.[82] Arsenic was, however, widely used for many medical, cosmetic and environmental applications during the nineteenth century, so he may have ingested it from some innocent source. Indeed, neutron activation analyses of his hair[77-80] found abnormal quantities of arsenic in some (but not all) samples from his period on St Helena. Lewin et al.[81] analysed a different sample of Napoleon's hair and found an almost normal arsenic content but elevated antimony, a component of many medications of that time, particularly tartar emetic. Recent studies by ICP-MS on a sample of Napoleon's hair revealed a relatively high concentration of 3.5 ppm As with a control value of 0.5 ppm As.[82] However, the reliability of both the source of the hair and the analytical result have been questioned by the US Federal Bureau of Investigation.

Analytes with Group IB–VIIB Elements

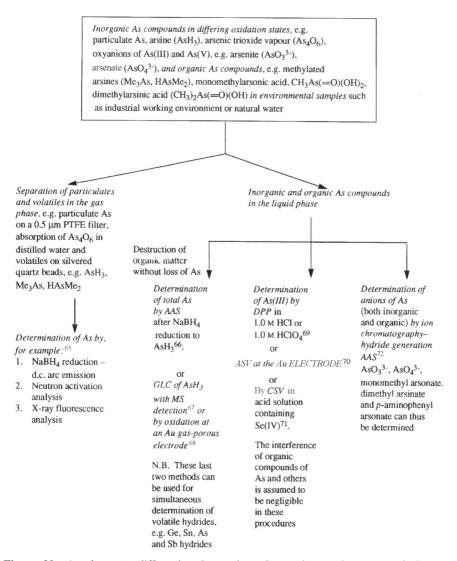

Figure 30 A scheme to differentiate inorganic and organic arsenic compounds in an environmental sample

4.7.3 Procedure

The method of Lewin et al.[81] was to place a 12 mg sample of the hair in a 1-ml polyethylene vial and irradiate it in a nuclear reactor for 16 h at a thermal neutron flux of 2.5×10^{11} n cm^{-2} s^{-1} for neutron activation analysis. After a 7-h delay, the hair was transferred to a clean polyethylene vial and re-weighed. The sample was then analysed for elements producing radioactive half-lives of 12.5–67 h (K, Na, As,

Br, Au and Sb) by counting the X-rays emitted from the sample with a 12.7% relative efficiency Ge–Li detector coupled to a hardwired 4196-channel multi-channel analyser. The nuclear reactor had the advantage of maintaining a reproducible and constant ($\pm 1\%$) neutron flux over the long term and the analytical system was calibrated previously with several standard reference materials, including NBS SRM 1632 Coal and SRM 1633 Fly Ash and with dilute solutions of arsenic salts. Stability of the specific activity produced by analysing various arsenic standards over nearly 6 years guaranteed the reliability of the calibration.

The sample was counted for 5–50 min to obtain various counting statistics. The concentrations of the elements K, Na, As, Br, Au and Sb in the hair were then calculated by comparing the net X-ray peak area with standard values. After preliminary counting, the hair was washed four times with absolute EtOH to remove surface contamination and was recounted several times over a period of a few days—the washing had no apparent effect on the elemental concentrations determined.

Table 26 shows that this sample of Napoleon's hair did not contain high levels of As (1.4 ± 0.2 ppm with control range 0.079–0.67 ppm) but did contain a moderately increased level of Sb (5.7 ± 0.6 ppm with control range 0.016–1.30 ppm).

Previous investigations had suggested that the maximum levels of As were located 4–9 cm from the proximal end of a 13-cm length of hair.[77] The average hair length of this sample, which measured 8 cm, was from the distal end and may not have recorded recent events. However, it would have contained high levels of As had there been chronic poisoning. Improved technology in the past 20 years compared with that used in the original investigations[77,84] has allowed clear separation of the radioisotopes of As, Sb and Br. The previous analytical method used[84] was not tested for the efficiency of separation of Sb and Br from As. Lack of such separation would tend to give apparent high arsenic levels. In addition, this 12-mg hair sample from Napoleon could be a more representative sample than the individual hair samples of less than 0.5 mg reported by Smith *et al.*[77]

Alternatively, during the nineteenth century, many people were accidentally poisoned by arsenical vapours from wallpaper.[85–87] The green copper arsenite pigments (Scheele's green and Paris or emerald green) were introduced in about 1780 and were soon widely used in paints and wallpapers. By 1815 they were known

Table 26 Elemental concentrations in Napoleon's hair. Reprinted with permission from *Nature*, 1982, **299**, 627. Copyright 1982 Macmillan Magazines Ltd

Element	Isotope	$t_{1/2}$ (h)	X-ray energy (keV)	Concentration (ppm)	
				Found	Control[83]
K	^{42}K	12.5	1,225	750 ± 100	0.94–310
Na	^{24}Na	15.0	1,369	$1,100 \pm 100$	0.4–850
Br	^{82}Br	35.5	777	4.2 ± 0.6	0.77–490
As	^{76}As	26.4	559	1.4 ± 0.2	0.079–0.67
Au	^{198}Au	64.7	412	0.45 ± 0.06	0.0006–1.36
Sb	^{122}Sb	67.1	564	5.7 ± 0.6	0.016–1.30

to be poisonous and throughout the nineteenth century many people were made ill or even killed by As from their wallpaper. Not until 1893 was the mechanism of the process elucidated.[88] On damp wallpaper, many moulds (e.g. *Scopulariopsis brevicaulis*) can metabolise arsenic compounds to volatile, poisonous arsine and/or arsenic trimethyl, which are released in the air of the room.

A sample of Napoleon's wallpaper had been preserved in an old family scrapbook.[89] It measured about 55 × 60 mm with a beige coating carrying a 40 mm diameter flock-decorated rosette, mainly brownish but with bright green areas. It had been removed from the drawing room of Longwood House (the room in which Napoleon died) and pasted into the scrapbook in May/June 1825. At the time, the room had a cream wallpaper with blue stars—the brownish rosette must be one of the stars, faded over the years.

This wallpaper and other samples from the nineteenth century were analysed primarily by X-ray fluorescence using a ^{147}Pm source of 1.5×10^9 Bq, an Si(Li) detector of area 80 mm^2 and a 2000-channel analyser covering the range 0–60 keV; all elements from K to Pb show either K or L X-ray fluorescence within this range. The spectra of the wallpapers were dominated by peaks due to Cu, As, Pb, Ca and Fe.

The spectra were calibrated by comparing them with those of the calibration papers made up to contain known amounts of these metals. Interpretation was complicated by the presence of Pb pigments in the papers. The Lα line of Pb almost coincides with the Kα line of As and the Pb components had to be subtracted from the recorded spectra before the As peaks could be measured. Furthermore, a dense layer of Pb pigment in the rosette of Napoleon's wallpaper attenuated the radiation of the metal pigments beneath it. They were estimated by reversing the paper in its mount and irradiating from the back. The final results are given in Table 27.

Matrix and attenuation effects, inevitable in the XRF analysis of heterogeneous samples, make the results uncertain by ±20–30%. Within these limits, the XRF results are supported by neutron and photon activation analyses of pigment flakes

Table 27 Metal content of historical wallpapers, determined by X-ray fluorescence spectroscopy. Reproduced with permission from *Nature*, 1982, **299**, 627. Copyright 1982 Macmillan Magazines Ltd

Element	X-ray observed	Metal content (as element) (g m^{-2})				
		Napoleon's wallpaper^a			Others	
		a	b	c	1	2
Cu	Kα	4.6	0.03	0.28	3.8	3.5
As	Kα	1.5	0.04	0.12	6.5	6.2
Pb	Lβ_1	19	0.3	1.3	1.9	0.4
Ca	Kα Kβ	6.6	14	14	7.5	17
Fe	Kα	0.15	0.4	0.4	0.07	0.06

^{a} a, Rosette area; b, beige area surrounding rosette; c, mean composition.

detached from Napoleons' wallpaper. The average composition of Napoleon's wallpaper was derived by taking the rosettes to be 12 cm^2 in area and spaced[90] at one per 225 cm^2 on the beige background. On this basis, the rosettes contribute 0.08 g m^{-2} of As and the beige background 0.04 g m^{-2}—a total of 0.12 g m^{-2}.

With this fairly mild As loading, Napoleon's wallpaper could not have posed a serious threat to life but it may well have contributed to his illness. In 1893,[87] a detailed study showed that wallpapers containing 0.6–0.15 g m^{-2} As could cause illness and suggested that even 0.006 g m^{-2} could be hazardous. The safety standards of the time[91] recommended 0.001–0.005 g m^{-2} as a safe upper limit. Thus a finding of 0.12 g m^{-2} of As in Napoleon's wallpaper is toxicologically significant, particularly in view of the As found in his hair by NAA. The studies, on hair samples spanning 1816–21, reported irregularly fluctuating levels of As, as would be expected from irregularly administered As medication or from a source such as damp and mouldy wallpaper, subjected to fluctuations of humidity, temperature, ventilation and the amount of time Napoleon spent in the relevant rooms.

This sample of Napoleon's wallpaper was only installed[90] in 1819, so cannot have contributed to his arsenic intake before that time. However, hazardously arsenical wallpaper was available on St Helena during Napoleon's captivity and, in one instance at least, used to decorate his residence. Accordingly, conspiracy theories need not be invoked to explain the As found in his hair.

4.8 DETERMINATION OF ANTIMONY IN LIVER SAMPLES USING HYDRIDE GENERATION–ATOMIC ABSORPTION SPECTROMETRY

4.8.1 Summary

Sb can be determined in liver samples by acid digestion, reduction of Sb(V) to Sb(III) followed by hydride generation–atomic absorption spectrometry with an LOD of 0.5 ng g^{-1}. Liver samples from deceased children who suffered the sudden infant death syndrome (SIDS) gave a mean [Sb] of 7.11 ng g^{-1} ($n = 37$) compared with a control group of $n = 15$ which gave a mean [Sb] of < 0.5 ng g^{-1}. It was suggested that SIDS children may have been exposed to a source of Sb within their environment.

4.8.2 Introduction

Recently, there has been the claim that antimony, used as a fire retardant by being added to the PVC coverings of baby cot mattresses, is a contributory factor in sudden infant death syndrome (SIDS). This has resulted in manufacturers withdrawing their mattresses from sale and bereaved parents have submitted their children's hair for analysis in an attempt to discover the cause of their children's deaths. Rooney (reported in Ref. 92) concluded that for every deceased child with a high level of antimony in its hair there was a corresponding high antimony level in its mattress. Levels of antimony in these cases would be of the order of 5 ppm with adult hair from the same families of the order of 300 ppb. The hypothesis for cot death is that fungal growth on these mattress coverings can release toxic gaseous compounds such as stibine, which can inhibit cholinesterase activity resulting in an accumulation of acetylcholine which leads to respiratory failure and death. Rooney[92] has also pointed out that antimony-loaded mattresses are not the only factor involved with cigarette smoking (60–70 ng Sb per cigarette), sleeping position of the infant, sex of the infant, etc., also being contributory. To investigate further the antimony factor, antimony concentrations in liver samples from SIDS and non-SIDS cases have recently been determined using hydride generation–atomic absorption spectrometry.[92]

4.8.3 Procedure

Liver specimens were collected at post-mortem and stored at $-80\,°C$. Prior to analysis, they were allowed to reach room temperature, weighed and transferred to 100 cm^3 acid-washed digestion tubes. The liver samples were dissolved with 5 cm^3 of HNO$_3$ in an aluminium heating block which was taken to 140 °C in 15 min and held at that temperature for a further 25 min. After cooling, 0.5 cm^3 of H$_2$SO$_4$ + 0.2 cm^3 of HClO$_4$ were added to the digestion completed by (i) heating to 140 °C in 15 min and the temperature held for 15 min, (ii) heating to 200 °C in 10 min and the temperature held for 15 min, (iii) heating to 250 °C in 10 min and the temperature held for 15 min, (iv) heating to 310 °C in 10 min and the temperature held for 20 min and (v) allowing to cool. Only the bottom 7 cm of the digestion tube was within the heating block, the upper portion acting as a condenser. At the completion of the heating process a small volume (ca 0.5 cm^3) of solution remained in the tube. A solution of 50% v/v HCl (5 cm^3) was then added and the tube was heated to 90 °C in 10 min and the temperature held for 20 min to reduce Sb(V) to Sb(III). The volume was made up to 10 cm^3 with water and the solution subjected to hydride generation–atomic absorption spectrometry. The reducing agent was 3% w/v NaBH$_4$ in 1% w/v NaOH. A reagent blank was subjected to this procedure with each batch of digestions. Calibration was performed using weighed portions of lamb liver to which Sb was added (1, 5, 10 and 20 ng) and these standards also subjected to acid digestion and reduction prior to hydride generation–atomic absorption spectrometry.

The reagent blank measurements ($n=8$) indicated the presence of very small amounts of antimony in the acids and set the limit of detection for the analytical method at 0.5 ng g^{-1}. Samples from 52 children were then analysed. These included 37 cases of SIDS and 15 non-SIDS controls. There was no measurable Sb in 14 of the non-SIDS specimens with the remaining sample in this group found to contain 1.65 ng g^{-1}. By contrast, Sb was found in 20 of the 37 SIDS specimens. The highest concentration was 111 ng g^{-1} with four samples containing more than 20 ng g^{-1} and 16 samples more than 2 ng g^{-1}. The mean [Sb] in the SIDS group was 7.11 ng g^{-1} and the mean [Sb] in the non-SIDS group was less than 0.5 ng g^{-1}. These groups were therefore significantly different with a p value of

0.002 using the Mann–Whitney test. These results therefore suggest that SIDS children may have been exposed to a source of Sb within their environment. Whether or not this source of Sb was directly responsible for causing SIDS is debatable, since such liver assays are indicative of chronic poisoning. Acute poisoning, more likely to cause SIDS, would be revealed by assay of blood or lung tissue.

4.9 DETERMINATION OF SELENIUM IN NATURAL WATERS BY SPECTROFLUORIMETRY

4.9.1 Summary

Se(IV) can be measured in natural waters by complexation with ammonium pyrrolidine dithiocarbamate at pH 4.2, extraction into $CHCl_3$, back-extraction into HNO_3, formation of the 2,3-diaminonaphthalene complex, solvent extraction into cyclohexane and spectrofluorimetric determination of the latter complex at 520 nm following excitation at 380 nm. This lengthy method allows the nanomolar determination of Se(IV) as opposed to total Se determination. By modification of this method, total Se concentrations can be measured and hence Se(VI) can be found by difference.

4.9.2 Introduction

Selenium originates from volcanic eruptions, in metallic sulphides such as those of Cu, Fe, Ni, Pb and in rare minerals such as Cu_2Se, $PbSe$ and As_2Se. It is seventeenth in crustal abundance and a biologically essential element for all vertebrates, albeit in trace amounts. Garlic and its relatives such as onions, leeks and chives have relatively high concentrations of selenium, which is believed to play a role in cancer prevention in animals and humans. Studies are currently in progress to understand the speciation of selenium in these plants, particularly when they are crushed, releasing volatile organoselenium compounds such as $R—S_x—Se_y—R'$.[93]

Its abundance in the environment is variable. For example, the Se content of soils ranges from 0.1 ppm in Se-deficient areas of New Zealand[94] to 1200 ppm in a region of Ireland,[95] at which concentrations it exercises toxic effects on animals that consume local vegetation. A wide concentration range is also found in non-saline waters from 0.1 μg dm^{-3} to extremes of 9 mg dm^{-3}.[96] Inorganic Se can exist in two oxidation states, Se(IV) (selenite or SeO_3^{2-}) and Se(VI) (selenate or SeO_4^{2-}) and, as a result of biological transformations, methylated species such as dimethyl selenide, dimethyl diselenide and dimethylselenone are also found.[97] These have been shown to be produced by microorganisms in sewage, sludge and soil[98] and are to be found in some natural waters. Selenite, selenate and organic selenide are the three distinct dissolved Se species in sea water[99] with the latter, believed to contain selenoamino acids and their derivatives, predominating in surface waters. In deep ocean waters selenite and selenate are the dominant species, with no evidence of organic selenide.

Selenium(IV) can be chelated with 2,3-diaminonaphthalene (DAN), which can be determined spectrofluorimetrically and this can be used for the determination of selenium(IV) in natural waters,[100] which is the subject of this section. Alternative techniques for the determination of Se are atomic absorption spectrometry,[101] flame photometry,[102] inductively coupled plasma spectrometry,[103] cathodic stripping voltammetry,[30] adsorptive stripping voltammetry,[104] gas chromatography[105] and coupled techniques such as gas chromatography–atomic fluorescence[106] and high-performance liquid chromatography–atomic emission spectrometry.[107]

4.9.3 Procedure and Calculation

(a) Sampling

A representative sample of say, river water would be acquired by taking 250-cm^3 samples from various positions at random and at various depths and then combining these to give a representative sample of 10 dm^3.

(b) Preliminary treatment of sample

The considerations that must be applied to the collection of samples for Se determination are similar to those for most other trace element work. Se is unusual in that it is necessary to consider volatile forms of the analyte. Once the representative sample has been obtained, particular regard must be paid to the possibility of biomethylation during storage and any losses by this route must be minimised. Quick freezing in liquid nitrogen, with samples stored in a freezer, is generally preferred as it minimises contamination and unforseen speciation changes. Pyrex glass can be used to store both acidified and unacidified water samples, e.g. selenite can be preserved in unacidified natural water samples for 2 weeks although 50% of selenite will be lost within 7 days by adsorption on the walls of the container.[108] Acidification can be helpful in the preservation of water samples, but this must not be overdone if speciation is to be maintained. Both Se(IV) and Se(VI) speciation is preserved by storing acidified (1 mol dm^{-3} H$_3$O$^+$) samples in Teflon-capped Pyrex glass.[109]

(c) Separation and preconcentration

Se(IV) is first complexed with ammonium pyrrolidine dithiocarbanate (APDC) at pH 4.2 and extracted into CHCl$_3$. This provides preconcentration of the analyte and separates Se(IV) from Se(VI). The selenium is then back-extracted from the CHCl$_3$ layer into nitric acid and the solution is evaporated to near dryness after adding a small quantity of perchloric acid.

(d) Determination step

The DAN complex is formed at pH 1, after first adding EDTA to complex other potentially interfering elements, followed by extraction into cyclohexane. The fluorescence is measured at 520 nm after excitation at 380 nm. The Se content is determined by the method of standard addition in which known amounts of Se are added to aliquots of the natural water sample which are then taken through the above procedure. The procedure for total Se involves precipitation as elemental Se by reduction with hydrazine sulphate in the presence of Te, which acts as a carrier. The Te is formed by a similar reduction of excess sodium tellurite and the precipitate is filtered off, dissolved in nitric acid containing a small amount of perchloric acid (70% w/v) and evaporated to near dryness. The DAN complex of Se is formed and its fluorescence is measured as before with Te not interfering in the measurement.

(e) Calculation step

The standard addition graph is given in Figure 31, i.e. a plot of fluorescence intensity vs concentration of added Se in nmol dm^{-3}.[110] Note that the fluorescence intensities are corrected by subtraction of appropriate blanks carried through both precipitation (total Se) and CHCl$_3$ extraction [Se(IV)] methods. The precipitation method gives a concentration of total Se of 2.56 nmol dm^{-3} (or 202 ng dm^{-3}) and the CHCl$_3$ extraction method gives a concentration of Se(IV) of 2.43 nmol dm^{-3} (or 192

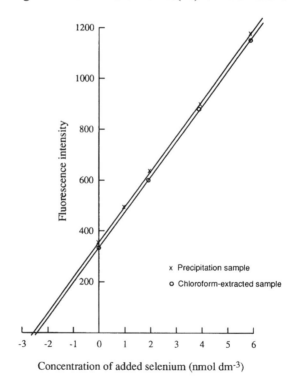

Figure 31 Calibration graphs for the determination of selenium by a standard addition method. Reproduced with permission

ng dm^{-3}); thus the difference, i.e. 10 ng dm^{-3}, relates to the concentration of Se(VI), but would depend on the precision of measurement of total Se and Se(IV) concentrations.

4.10 DETERMINATION OF FLUORIDE IN POTABLE WATER BY ION-SELECTIVE ELECTRODE MEASUREMENTS

4.10.1 Summary

F$^-$ can be measured directly in potable waters at ppm and sub-ppm concentrations by an ion-selective electrode. Using a calibration graph or standard addition method and a TISAB reagent, the method is rapid and measures the free F$^-$ species with relative standard deviations of a few per cent. The method can be adapted for F$^-$ measurement in other complex matrices such as toothpaste and soil, requiring preliminary sample treatment steps.

4.10.2 Introduction

The success of the glass electrode for pH measurement led to many attempts to develop electrodes for various other ions. The so called 'specific-ion' electrodes first appeared in the 1960s with the Ca^{2+}-selective electrode in 1966 (liquid membrane type) and the F$^-$-selective electrode in 1967 (solid-state type). The latter electrode was the first non-glass membrane solid-state electrode and was fabricated using an LaF$_3$ membrane which interfaced an internal reference electrode of Ag/AgCl dipped in a solution containing Na$^+$, Cl$^-$ and F$^-$ ions with the external solution to be measured for [F$^-$]. Conduction through the membrane is facilitated by the movement of F$^-$ ions between anionic lattice sites, which is in turn influenced by the F$^-$ activities on either side of the membrane.

The potential of a cell, E_{cell}, comprising the F$^-$-selective electrode and an external reference electrode, both dipped in a solution of unknown [F$^-$], is given by

$$E_{cell} = k - \frac{0.059}{1} \times \log_{10}[F^-] \tag{32}$$

This Nernstian response is given by the F$^-$ electrode over the concentration range 1–10^{-6} mol dm^{-3} with a selectivity constant K_{F^-,A^-} of <0.001 for all other anions except OH$^-$, for which it is of the order of 10^4. Thus,

$$E_{cell} = k - \frac{0.059}{1} \log\{[F^-] + k_{F^-,A^-}[A^-]^{1/z}\} \tag{33}$$

where z is the charge on the interfering ion. The optimum pH range for usage of this electrode is 5–8.

The importance of this electrode in F$^-$ measurement in aquatic chemistry, industrial chemistry, biological studies, etc., is illustrated in Figure 32.

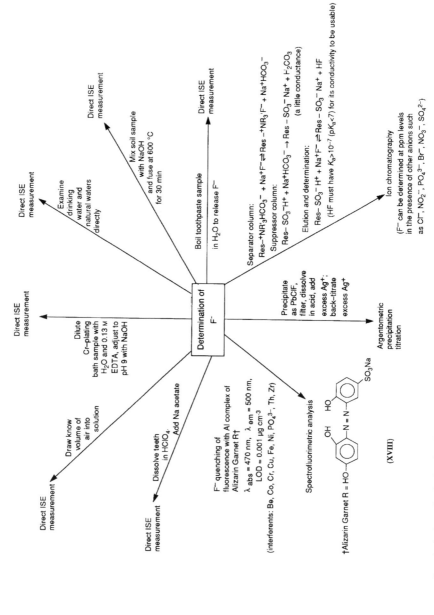

Figure 32 Summary of analytical methods for the determination of [F$^-$] in selected complex matrices

4.10.3 Procedure

F^- ions are in some cases added to drinking water at concentrations of the order of 1 ppm (i.e. 1 mg dm^{-3}) to help prevent tooth decay. The F^- ion-selective electrode in conjunction with an Ag/AgCl reference electrode can therefore be used to check that the correct quantity of F^- has been added. Such drinking water samples would also be expected to contain the ions Na^+, K^+, Mg^{2+}, Ca^{2+}, Al^{3+}, Fe^{2+}, Fe^{3+}, Cl^- and NO_3^-. Determination of unknown $[F^-]$ can be made by the use of a calibration graph (constructed daily), the standard addition method of quantitation or potentiometric titration. Using the former method, calibration standards could range from 0.1 to 10 ppm and an ionic strength adjuster (ISA) should be added to each standard and sample. In the case of $[F^-]$ determination, a total ionic strength adjuster buffer (TISAB)...

(a) gives a constant ionic strength to allow the Nernst equation to be written with concentration rather than activity values (i.e. TISAB is relatively concentrated in NaCl to give essentially the same total ionic strength in both standards and sample and hence a constant activity coefficient);
(b) contains a complexing agent, 1,2-diaminocyclohexanetetraacetic acid (CDTA), which chelates Al^{3+} and other metal ions that could complex with F^- (note: the electrode only responds to free F^- and not to AlF_6^{3-} in the case of Al^{3+}); and
(c) also contains a buffer to ensure that the electrode operates in the optimum pH range 5–8.

Orion recommends the preparation of TISAB as follows. Add 57 cm^3 of glacial ethanoic acid, 58 g of NaCl and 4 g of CDTA to 500 cm^3 of distilled water in a 1-dm^3 beaker. Stir to dissolve and cool to room temperature since the slope of calibration graph is dependent on temperature, i.e. $2.303RT/nF$. Adjust the pH to between 5 and 5.5 using 5 mol dm^{-3} NaOH. Pour the mixture into a 1-dm^3 volumetric flask and make up to the mark.

Each standard and sample solution (25 cm^3) should be diluted with an equal volume of TISAB and allowed to equilibrate to room temperature. The cell emf of each of these solutions should then be measured with the F^--selective electrode–Ag/AgCl reference electrode combination. A graph of emf vs concentration should be plotted for the standards using semi-logarithmic graph paper. The unknown $[F^-]$ of each sample can then be read off from the calibration plot as shown in Figure 33 (see Table 28).

4.10.4 Calculation

A 3.045-g sample of toothpaste was suspended in 50 cm^3 of fluoride-strength buffering medium (TISAB) and boiled briefly to extract the fluoride. The mixture was cooled, transferred quantitatively into a 100-cm^3 volumetric flask and diluted to volume with deionised water. A 25-cm^3 aliquot was transferred into a beaker, a fluoride ion-selective electrode and reference electrode were inserted and a potential of -155.3 mV was obtained after equilibration. A 0.10-cm^3 aliquot of 0.5 mg cm^{-3} fluoride stock solution was added, after which the potential was -176.2 mV. Calculate the percentage of fluoride by weight in the original toothpaste sample.

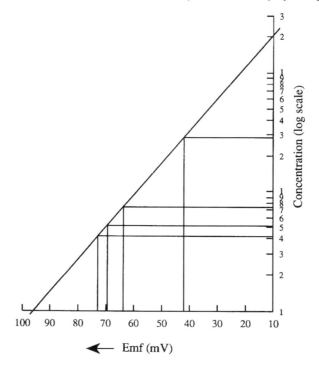

Figure 33 Calibration plot using a fluoride ion-selective electrode. Reproduced with permission

Table 28 Calibration data for F^- ion-selective electrode and determination of unknowns

Sample	Concentration of F^- (ppm)	Emf (mV)
Standards	1.00	96.0
	5.00	70.5
	10.00	59.0
	50.00	32.5
	100.00	21.5
Unknowns	4.2	73.0
	5.2	69.5
	28	42.0
	7.4	63.5

Using equation 32 for the unknown:

$$-0.1553 = k - \frac{0.059}{1} \log[F^-] \qquad (34)$$

Using equation 32 for the unknown + standard addition:

$$-0.1762 = k - \frac{0.059}{1} \log\left([F^-] \times \frac{25}{25.1} + \frac{0.1 \times 0.5}{25.1}\right) \qquad (35)$$

Subtract equation 34 from equation 35:

$$0.0209 = 0.059 \log\left(\frac{[F^-] \times \frac{25}{25.1} + \frac{0.1 \times 0.5}{25.1}}{[F^-]}\right)$$

$$0.0209 = 0.059 \log\left(\frac{25}{25.1} + \frac{0.1 \times 0.5}{25.1[F^-]}\right)$$

$$0.353 = \log\left(0.996 + \frac{1}{502[F^-]}\right)$$

$$0.996 + \frac{1}{502[F^-]} = 2.255$$

$$\frac{1}{502[F^-]} = 1.259$$

$$[F^-] = \frac{1}{1.259 \times 502} \text{ mg cm}^{-3}$$

$$\text{Fluoride (\%)} = \frac{1}{1.259 \times 502} \times \frac{1}{10^3} \times \frac{100}{1} \times \frac{1}{3.045} \times 100$$

$$= 0.0052$$

4.11 MULTIELEMENT ANALYSIS OF A BIOLOGICAL FLUID BY INDUCTIVELY COUPLED PLASMA MASS SPECTROMETRY (ICP-MS)

4.11.1 Summary

Inductively coupled plasma mass spectrometry can be used to identify and determine rapidly a wide variety of elements in complex matrices such as a whole blood sample. The identification of a particular element is based on the *m/z* value that it exhibits in the mass spectrum together the mass spectral pattern of the element's natural isotopes. Determination is possible in the range 1–100 ppt. The technique is expensive, particularly for automated, simultaneous, multielement analysis, but would be regarded as an analytically powerful technique in elemental mapping of complex matrices of biomedical and environmental significance. In this particular case it is used to indicate occupational exposure of a worker to Pb, Ba and Sn.

4.11.2 Introduction

Simultaneous multielement analysis is increasingly being carried out by inductively coupled plasma atomic emission spectrometry (ICP-AES) and inductively coupled plasma mass spectrometry (ICP-MS). The main advantage of a plasma source relative to the familiar flame source used in atomic absorption spectrometry (AAS) is the increased temperature which can be achieved. While N_2O–C_2H_2 in AAS can achieve a maximum temperature of 3000 K, the ICP source has exhibited temperatures in the range 5000–10 000 K. At these higher temperatures, most atoms will emit strongly at the resonance wavelength for multielement analysis in ICP-AES and most elements are also significantly ionised, allowing ICP to be an ion source for multielement analysis by mass spectrometry. In the latter case, the different ions are separated in the mass spectrometer according to their m/z ratios and are determined using calibration curves constructed from plots of ion count at the detector vs concentration. ICP-MS can perform rapid automated multielement qualitative analysis since the mass spectral patterns of individual elements are unique and relatively simple, e.g. for Se the natural isotopic abundances are ^{74}Se 0.9, ^{76}Se 9.0, ^{77}Se 7.6, ^{78}Se 23.6, ^{80}Se 49.8 and ^{82}Se 9.2; these relative abundances are then observed in the mass spectrum in terms of counts vs atomic mass and thus provide incontestable evidence for the presence of Se in the particular sample. ICP-AES can also perform a limited qualitative survey of a sample, but the lack of accurate data on relative line intensities and spectral interferences render it less powerful as a qualitative tool than ICP-MS. Flame AAS and graphite furnace GF AAS are both theoretically capable of performing qualitative analysis but coverage of a large range of elements is impractical owing to the single-element nature of the techniques.

ICP-MS is superior to GFAAS, flame AAS and ICP-AES in terms of LODs for elemental determination in solution, with LODs in the range 1–100 ppt. From the results given in Section 4.11.3 it can perform quantitative analysis on a complex sample with accuracies and precisions comparable to those of other analytical techniques. However, interference effects can occur in ICP-MS. For example, below mass 80 major background species arise from the argon isotopes (from the plasma gas), from the major elements of the solvent (e.g. oxygen and hydrogen) and from the acid used to prepare the sample (e.g. nitrogen from nitric acid, chlorine from hydrochloric acid, sulphur from sulphuric acid). The sample matrix can also contribute significantly to spectral interference in ICP-MS. Molecular ions, particularly oxides and hydroxides, are the most important interferents, e.g. $^{35}Cl^{16}O^+$ causes a severe interference on $^{51}V^+$, the vanadium isotope that is 99.76% abundant. The determination of As in sea water is also complicated by interference due to $^{40}Ar^{35}Cl$ on ^{75}As. In general, despite these and other interferences, excellent analytical results can be achieved by ICP-MS provided that recommended instrument optimisation is performed, the sample preparation methodology takes account of possible spectral interferences from acids used for dissolution and internal standards are chosen to compensate for matrix-induced suppressions and enhancements of analyte signals.

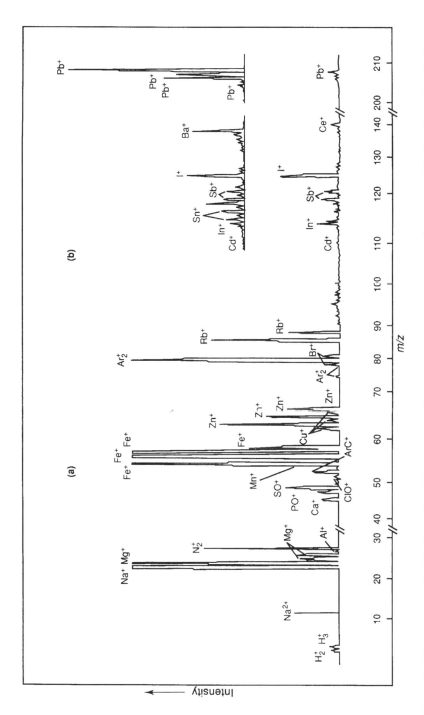

Figure 34 Mass spectra of whole blood from (a) a woman with no excessive exposure to trace elements and (b) from her husband who was occupationally exposed to lead. Reproduced by permission of the Royal Society of Chemistry from *Chem. Br.*, October 1988, 1009

Quadrupole-based ICP-MS is a quasi-simultaneous multielement technique in that the complete element range can be scanned and approximately 100 spectra averaged per minute. Autosamplers can be incorporated in ICP-MS to help increase the sample throughout. ICP-AES and ICP-MS systems for simultaneous, multielement analysis are, however, the most expensive instruments to purchase compared with those for sequential ICP-AES, flame AAS and GFAAS. The operating costs of ICP instruments are mainly due to the cost of the argon used as the main plasma gas.

4.11.3 Procedure

The rapid determination of the concentrations of a wide variety of elements in a whole blood sample is relatively straightforward by ICP-MS, e.g. a 1000-μl sample of whole blood is diluted $1 + 24$ with 1% v/v of Triton X-100 (polyethylene glycol alkyl aryl ether) and this is then introduced into the plasma by pneumatic nebulisation (spraying) prior to detection of the resulting ions by a quadrupole mass spectrometer.[111] In this particular case quantitative data were obtained for at least 15 elements in just 10 min. Figure 34a shows the mass spectrum of a blood sample from a women with no excessive exposure to any trace element and, for comparison, Figure 34b shows the partial mass spectrum of blood from her husband who had occupationally been exposed to lead.[111] The only substantial differences in the spectra occurred above 110 amu, showing clearly the 20-fold increase in blood lead for the husband (4 vs 0.2 μmol dm^{-3}) plus evidence of his increased exposure to Ba and Sn at ca 50 and 70 nmol dm^{-3}, respectively. Iodine, an essential trace element, is present at equal concentrations of ca 100 nmol dm^{-3} in both specimens. The lower mass region of the spectrum contains peaks from Na, Ca, Mg, Fe and P at 10–150 mmol dm^{-3}, Cu, Zn, Rb and Br at 10–100 μmol dm^{-3} and Al and Mn at below 1 μmol dm^{-3}.

Chapter 5

Selected Analytical Problems Involving Organic Analytes which are Major or Minor Organic Constituents of a Sample

5.1 COMPARISON OF ANALYTICAL METHODS BASED ON VISIBLE SPECTROPHOTOMETRY, SOLVENT EXTRACTION–UV SPECTROPHOTOMETRY, VOLTAMMETRY, NMR AND HPLC FOR THE DETERMINATION OF THE ACTIVE CONSTITUENTS OF ANALGESIC FORMULATIONS

5.1.1 Summary

Aspirin (acetyl-o-salicylic acid) can be selectively determined in a pharmaceutical formulation by chelation of its hydrolysis product, o-salicylic acid, with Fe^{3+}. Other constituents of analgesic formulations such as paracetamol, caffeine and phenacetin will not give this reaction and hence do not interfere. The method is, however, time consuming and is ideally used when alternative instrumentation is unavailable. Other analytical methods display different selectivities. APC tablets, which contain aspirin, phenacetin and caffeine, can be assayed by solvent extraction–UV spectrophotometry by a relatively time-consuming method which involves the separation of aspirin from the other two constituents of the formulation prior to their UV spectrophotometric determination. The electrochemical oxidation of paracetamol and its degradation product (p-aminophenol) leads to a voltammetric method for their rapid and selective determination, since other analgesics do not interfere at the potentials of oxidation. NMR spectrometry involves the use of sophisticated instrumentation for the rapid and specific assay for a particular analgesic such as aspirin in a multianalgesic formulation. The optimum selectivity for the determination of multianalgesics can be obtained using HPLC, by means of which most of the analgesics can be determined rapidly in a single run. Given the availability of HPLC instrumentation in the laboratory, this would be regarded as the method of choice for the analysis of multianalgesic formulations.

5.1.2 Visible Spectrophotometric Method

5.1.2.1 INTRODUCTION

Aspirin (**XIX**) in analgesic formulations can be acid hydrolysed to give *o*-salicylic acid (**XX**). This will chelate with Fe^{3+} to give a violet species of octahedral geometry (**XXI**). Its colour arises from electronic transitions between σ or Π ligand orbitals and empty or antibonding metal orbitals.

(**XIX**) (**XX**) (**XXI**)

Chelation of **XX** with Fe^{3+} endows this analytical method with a particular specificity in the event of the formulation containing other analgesics which do not chelate with Fe^{3+}, e.g. phenacetin (**XXII**), caffeine (**XXIII**) and paracetamol (**XXIV**). The method has also the advantage that it uses commonplace laboratory instrumentation, but it is time consuming.

(**XXII**) (**XXIII**) (**XXIV**)

5.1.2.2 PROCEDURE

Hydrolysis of the analgesic formulation containing the aspirin is first carried out in dilute sulphuric acid to give the parent salicylic acid (**XX**). This 2-hydroxybenzoic acid, on mixing with Fe^{3+} solution, gives a violet complex. The coloration thus produced is stable and the concentration of the complex is then found spectrophotometrically via a calibration graph. It has been found better to use the sodium salt of the salicylic acid to give the calibration graph because of its improved solubility compared with the parent acid. The strength of this solution of 5.78 g of sodium 2-hydroxybenzoic acid in 1 dm^3 of distilled, deionised water is chosen so that each 1 cm^3 is equivalent to 5 mg of 2-hydroxybenzoic acid.

Organic Analytes Which are Major or Minor Sample Constituents

Table 29 Preparation of calibration standards for aspirin determination by visible spectrophotometry

Solution No.	Sodium 2-hydroxybenzoate (vol./cm^3) (V_s)	Water (vol./cm^3)	Dilute sulphuric acid[a] (vol./cm^3)
1	1	9	18
2	2	8	18
3	3	7	18
4	4	6	18
5	5	5	18
6	6	4	18
7	0	10	18

[a] 15 cm^3 of concentrated H$_2$SO$_4$ made up to 1 dm^3 with distilled water.

(a) Calibration graph

A series of solutions as indicated in Table 29 is prepared in boiling tubes. Warm these tubes in a boiling water-bath for ca 15 min, cool and add 2 cm^3 of Fe^{3+} solution (0.6 mol dm^{-3} FeCl$_3$) to each tube. Using a UV visible spectrophotometer and 1-cm path length cells, record the UV–visible spectrum of calibration standards Nos 1–6 over the range ca 370–700 nm, with solution No. 7 containing no salicylate as reference in the back beam of the instrument. From the observed spectra, plot the baseline-corrected absorbance at λ_{max} 510 nm vs concentration to obtain the calibration plot from the standards Nos 1–6.

(b) Sample treatment

One pre-weighed tablet of aspirin is dissolved in 200 cm^3 distilled water and is heated to near boiling for 15 min to ensure complete dissolution of aspirin. Any residue is ignored at this stage. Then 10 cm^3 of the cooled aspirin solution are pipetted into a boiling tube, 18 cm^3 of the dilute sulphuric acid are added and boiled a further 15 min, filtering off any residue. The resultant solution is cooled and 2 cm^3 of the Fe^{3+} solution are added. The visible spectrum is recorded as for the calibration samples and the absorbance at λ_{max} 510 nm is read off.

5.1.2.3 CALCULATION

From the calibration graph, estimate for the sample the corresponding volume V_s of sodium 2-hydroxybenzoate equivalent. The solution on which the absorbance measurement was made (30 cm^3 total volume) thus contains $5 V_s$ mg of 2-hydroxybenzoic acid. Hence the aspirin tablet contains the equivalent (as the acid) of $5 V_s \times 200/10$ mg. Hence the weight of aspirin in the original formulation can be calculated as

$$5V_s \times \frac{200}{10} \times \frac{180}{138} \text{ mg}$$

where **XIX** has a molecular weight of 180 and **XX** has a molecular weight of 138.

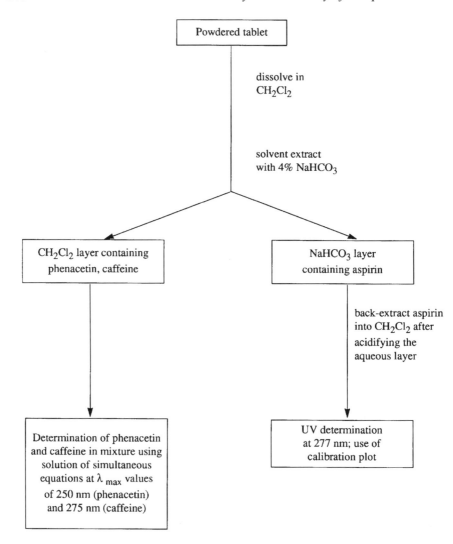

Figure 35 Determination of aspirin, phenacetin and caffeine in APC tablets by solvent extraction–UV spectrophotometry

5.1.3 Solvent Extraction–UV Spectrophotometric Method

5.1.3.1 INTRODUCTION AND PROCEDURE

APC tablets, which contain aspirin (**XIX**), phenacetin (**XXII**) and caffeine (**XXIII**), can be determined by the scheme shown in Figure 35.

The method is again relatively time consuming, is not specific for aspirin (**XIX**), requiring solvent extraction from an alkaline solution to partition the weak base caffeine (**XXIII**), pK_a 0.6) and phenacetin (**XXII**), a neutral molecule at mildly alkaline pH, into dichloromethane, leaving aspirin as the carboxylate anion in the aqueous phase. This method again possesses the advantage of using readily available instrumentation in the form of a UV–visible spectrophotometer.

5.1.4 Voltammetric Method

5.1.4.1 INTRODUCTION

Shearer[112] reported the determination of the major degradation product (*p*-aminophenol) in formulations containing the electrooxidisable acetaminophen (paracetamol) (**XXIV**).

5.1.4.2 PROCEDURE

A solution containing 50 mg of **XXIV** in 250 cm^3 of methanol was mixed with an equal volume of 0.1 mol dm^{-3} sodium acetate in methanolic 0.1 mol dm^{-3} acetic acid, deoxygenated with nitrogen and a voltammogram obtained between -0.2 and $+0.6$ V. Under these conditions, *p*-aminophenol is oxidised in a 2e$^-$ step to quinoneimine with a peak potential of $+0.2$ V. Acetaminophen (**XXIV**) oxidises at $+0.5$ V. At a molar fraction of 0.92 of acetaminophen the relative error in voltammetric determination was found to an acceptable $+1\%$. The peak height of this standard was then compared with that of a sample taken through the above procedure and the paracetamol content calculated accordingly.

It was also found that the presence of water in methanolic solutions makes little difference in the results up to 10%. Caffeine (**XXIII**), aspirin (**XIX**), codeine phosphate, salicylamide (**XXV**), amphetamine sulphate, pentobarbital, prednisolone and other excipients were found not to interfere in the determination. Ascorbic acid was the only interferent manifested since it gives rise to an oxidation peak at the glassy carbon electrode. This method is therefore very selective for paracetamol (**XXIV**) and its degradation product (*p*-aminophenol), is relatively rapid, but uses instrumentation not always available in analytical laboratories.

5.1.5 NMR Method

5.1.5.1 INTRODUCTION

Aspirin can be determined in mixtures with phenacetin and caffeine by NMR spectrometry[113] by virtue of the fact that the proton resonance for CH$_3$CO in aspirin occurs at a different δ value to proton resonances for phenacetin and caffeine (Figure 36).

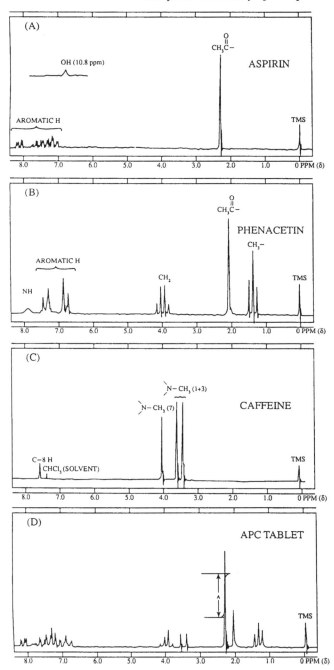

Figure 36 NMR spectra of aspirin (A), phenacetin (B), caffeine (C) and a mixture of the three (D). Reprinted with permission from D. P. Hollis, *Anal. Chem.*, 1963, **35**, 1682. Copyright 1963 American Chemical Society

This analytical method uses sophisticated instrumentation and is specific for molecules such as aspirin, which has a proton resonance that does not overlap with proton resonances of other analgesics in the sample. It is, however, a rapid method of assay.

5.1.5.2 CALCULATION

X mg of aspirin in 60 mg of sample dissolved in 0.5 cm^3 of CDCl$_3$, i.e. $(X/\text{MW}_{\text{asp}}) \times 10^{-3}$ mol, gives an integrated signal of A units at 2.3 δ, where MW$_{\text{asp}}$ is the molecular weight of aspirin.

Using caffeine and the N-CH$_3$ (position 7) proton resonance at 4 δ as a standard which produces an integrated signal I_{caff} for 25 mg of caffeine dissolved in 0.5 cm^3 of CDCl$_3$, i.e. $(25/\text{MW}_{\text{caff}}) \times 10^{-3}$ mol, where MW$_{\text{caff}}$ is the molecular weight of caffeine. Therefore,

$$1 \text{ mol gives a signal of } \frac{I_{\text{caff}} \times \text{MW}_{\text{caff}}}{25} \times 10^3$$

$$\frac{X}{\text{MW}_{\text{asp}}} \times 10^{-3} \text{ mol gives } \frac{I_{\text{caff}} \times \text{MW}_{\text{caff}}}{25} \times 10^3 \times \frac{X}{\text{MW}_{\text{asp}}} \times 10^{-3}$$

which is equivalent to signal A units at 2.3 δ. Therefore,

$$\text{mg aspirin} = X = \frac{A \times 25 \times \text{MW}_{\text{asp}}}{I_{\text{caff}} \times \text{MW}_{\text{caff}}}$$

$$\text{mg aspirin/mg sample} = \frac{A \times 25 \times \text{MW}_{\text{asp}}}{I_{\text{caff}} \times \text{MW}_{\text{caff}} \times 60}$$

It should be noted that both signal A and I_{caff} involve three protons each.

5.1.6 HPLC Method

5.1.6.1 INTRODUCTION

The optimum selectivity for the determination of multianalgesic formulations can be obtained using HPLC.[114] It was possible to achieve the simultaneous quantitation of **XXIV, XIX, XXIII**, codeine phosphate, **XXII, XX** and salicylamide (**XXV**).

(XXV)

Table 30 HPLC assay of a multianalgesic formulation

Ingredient	Claimed per tablet (mg)	Found (% of claim)
Acetaminophen	97.2	98.6
Aspirin	194.4	99.6
Salicylamide	129.6	100.0
Caffeine	64.8	99.1

5.1.6.2 Procedure

The sample is dissolved in EtOH, diluted with H_2O and an aliquot injected on to a reversed-phase HPLC column (μBondapak C_{18}) with 10 mmol dm^{-3} of KH_2PO_4–MeOH (81 + 19) adjusted to pH 2.3 with H_3PO_4 as the mobile phase and a detection wavelength of 254 nm. Lowering the pH to 2.3 increased the retention time of some weak acids and decreased that of some weak bases and hence made the above separation possible. Calculation of the analgesic content of the formulation was effected using the internal standard method. This analytical method gave excellent assay results for a commercial product (e.g. Table 30) and, given the availability of HPLC instrumentation in the laboratory, would be regarded as the method of choice for such multianalgesic formulations.

5.2 STABILITY-INDICATING HIGH-PERFORMANCE LIQUID CHROMATOGRAPHIC ASSAY FOR OXAZEPAM TABLETS AND CAPSULES

5.2.1 Summary

A stability-indicating HPLC assay for the 1,4-benzodiazepine oxazepam in capsules and tablets can be carried out by extraction of the sample with methanol–water (95 + 5), followed by chromatography on a C_{18} reversed-phase column with methanol–water–acetic acid (60 + 40 + 1) as mobile phase and a detection wavelength of 254 nm. This method separates oxazepam from all degradation products mentioned in the literature or observed in stress-degraded samples. Degradation products can be detected at the 0.1% level.

5.2.2 Introduction

Various analytical methods have been described for the determination of oxazepam (**XXVI**), a 1,4-benzodiazepine tranquilliser. An official method[115] employs UV spectrophotometric determination after extraction with ethanol. Increased selectivity can be observed using polarographic detection, which is dependent on reduction of the intact 4,5-azomethine bond. During GLC analysis, oxazepam decomposes to 6-chloro-4-phenyl-2-quinazolinecarboxaldehyde (**XXVII**). GLC methods, therefore, have been based on measurement of either **XXVII** or 2-amino-5-chlorobenzophenone (**XXVIII**), which is formed by prior hydrolysis of oxazepam. A highly selective high-performance liquid chromatographic method can be used for dosage forms in the presence of degradation products[116] such as **XXVII–XXXV** (see Figure 37). These degradation products, either mentioned in the literature or observed in stress-degraded samples, could be detected at the 0.1% level.

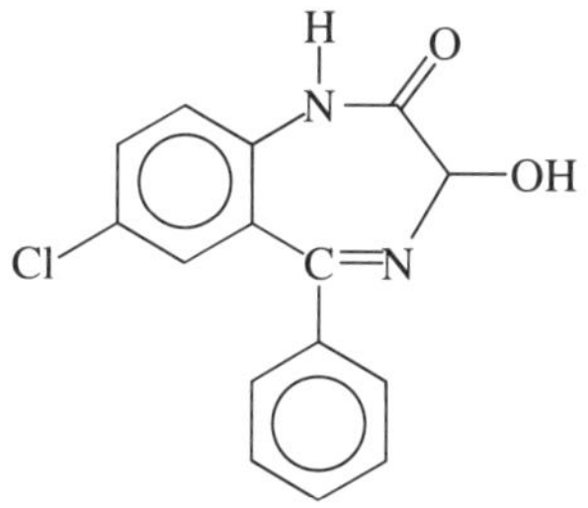

(**XXVI**)

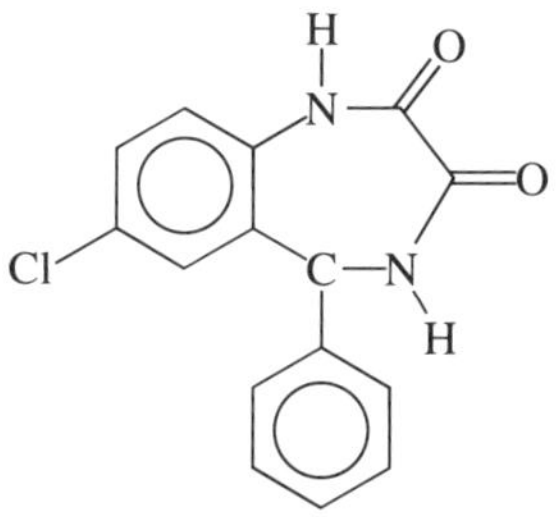

(**XXXI**)

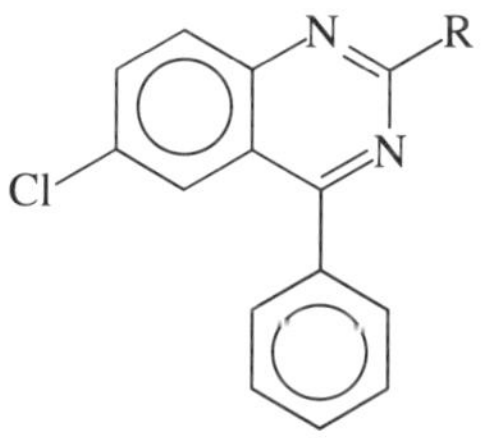

(**XXVII**) R = CHO
(**XXXV**) R = CH$_2$OH

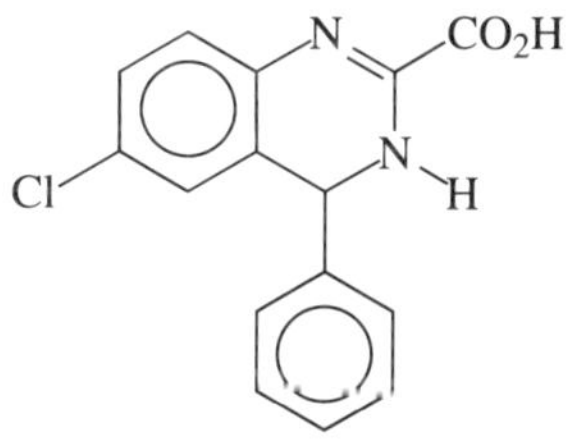

(**XXXII**)

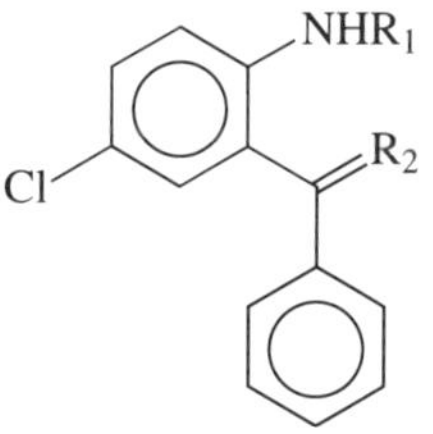

(**XXVIII**) R$_1$ = H, R$_2$ = O
(**XXXIII**) R$_1$ = COCHO, R$_2$ = O
(**XXIX**) R$_1$ = COCHOHNH$_2$, R$_2$ = O
(**XXX**) R$_1$ = H, R$_2$ = NHCHOHCO$_2$H

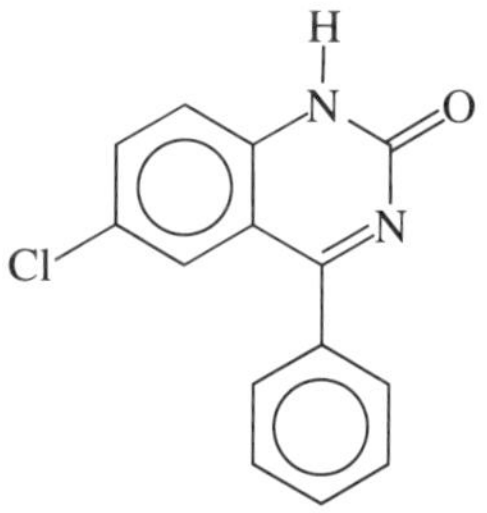

(**XXXIV**)

Reproduced by permission of the American Pharmaceutical Association from
J. Pharm. Sci., 1983, **72**, 1330

5.2.3 Procedure

The chromatograph was equipped with a UV detector set at 254 nm. A 10-μl loop was used in conjunction with 10-μm microparticulate reversed-phase columns of 25–30 cm × 4.6 mm i.d. The mobile phase consisted of methanol–water–acetic acid (60 + 40 + 1) with a flow rate of 2 cm^3 min^{-1}. An electronic integrator was used for area determinations; peak heights were determined manually.

One of the criteria for analytical method validation and system suitability testing, resolution R (equation 7), was tested by dissolving ca 10 mg of oxazepam and 15 mg of the main degradation product in stress-degraded samples (**XXVII**) in 250 cm^3 of methanol; 10 μl of this solution were chromatographed using the assay conditions. The resolution R between oxazepam and **XXVII** is ≥ 5.0. Unsuitable resolution can often be improved by a slight reduction of the methanol concentration in the mobile phase.

The standard solution was prepared by dissolving a quantity of oxazepam standard in methanol–water (49 + 1) to give a known concentration of ca 0.1 mg cm^{-3} and subjected to HPLC. Samples were prepared by emptying ≥ 20 capsules and determining the average weight, or by weighing ≥ 20 finely powered tablets. An accurately weighed portion of the powder containing ca 25 mg of oxazepam was transferred in a 250-cm^3 volumetric flask, 5 cm^3 of water were added

Table 31 Method reproducibility and recovery for the HPLC assay of oxazepam in capsules and tablets. Reproduced by permission of the American Pharmaceutical Association from Reif et al., *J. Pharm. Sci.*, 1983, **72**, 1330

Oxazepam product	Precision[a]
10-mg Capsule	98 ± 0.9 (height)
	98 ± 1.3 (area)
30-mg Capsule	97 ± 1.9 (height)
	97 ± 1.9 (area)
15-mg Tablet	99 ± 0.72 (height)
	99 ± 0.35 (area)

	Recovery (%)	
Composition[b]	10 parts oxazepam	30 parts oxazepam
A	99	101
B	99	101
C	101	99

[a] Average percentage of label claim ± *RSD* ($n = 5$).
[b] (A) lactose–Ac Di Sol–stearic acid–methylcellulose (300 + 6 + 3 + 60) and oxazepam (10 or 30); (B) dicalcium phosphate dihydrate–polacrilin potassium–starch–magnesium stearate (300 + 6 + 6 + 3) and oxazepam (10 or 30); (C) Avicel–alginic acid–sodium starch glycolate–talc (300 + 6 + 6 + 3) and oxazepam (10 or 30).

and the flask was swirled to wet the powder. Approximately 200 cm^3 of methanol were added, the solution was sonicated for 5 min, stirred for 30 min and diluted to volume with methanol. A portion was centrifuged to provide a clear solution for injection into the chromatograph. The ratio of the peak areas/heights of standard and sample solutions was used to evaluate the oxazepam content of formulations.

The average percentage of the label claim ± the relative standard deviation ($n = 5$) obtained by replicate assays of commercial samples and recovery data for synthetic mixtures of **XXVI** and three combinations of commonly used excipients are reported in Table 31. Placebo interference due to each mixture was stated to be < 0.1% for the method. Linearity of response in the range 0.06–0.12 mg cm^{-3} was observed in the assay of solutions.

HPLC separated **XXVI** from all of the degradation products reported in the literature and observed in stress-degraded samples. A separation of oxazepam from these degradation products at levels of ca 0.5% is shown in Figure 37.

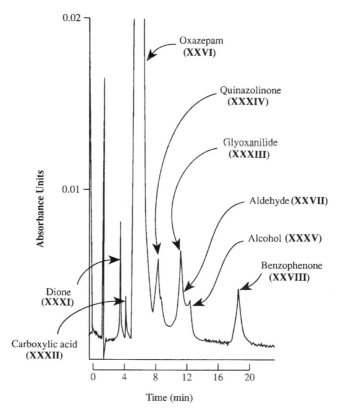

Figure 37 HPLC of a synthetic mixture of oxazepam and its degradation products at levels of ca 0.5%. Reproduced by permission of the American Pharmaceutical Association from Reif et al., *J. Pharm. Sci.*, 1983, **72**, 1330

5.3 SEPARATION OF WATER-SOLUBLE VITAMINS BY CAPILLARY ZONE ELECTROPHORESIS AND MICELLAR ELECTROPHORETIC CAPILLARY CHROMATOGRAPHY (MECC)

5.3.1 Summary

Water-soluble vitamins such as B_1, B_2, B_3 and B_6 can be separated with high efficiency by capillary zone electrophoresis (CZE) and migrate according to their relative charges. Micellar electrophoretic capillary chromatography (MECC) alters the order of migration, which depends on the extent of binding of particular vitamins within the lipophilic core of the micelles. Either CZE or MECC can be used to determine these vitamins in vitamin B complex formulations by a simple method. This involves dissolution of the vitamins in the formulation in an appropriate solvent, filtering a portion of the supernatant through a 0.45-μm filter into a CE sample vial followed by CZE or MECC using a background electrolyte of sodium tetraborate–sodium dihydrogen phosphate (20 mmol dm^{-3}, pH 9), various concentrations of sodium dodecyl sulphate (0, 50, 100, 150 mmol dm^{-3}) and a detection wavelength of 200 nm.

5.3.2 Introduction

The technique and applications of capillary electrophoresis have received great attention in recent years. In addition to the separation and determination of biomolecules such as proteins and nucleotides, the technique is now successfully used in environmental analysis, drug analysis in therapeutic and forensic situations, etc. Capillary zone electrophoresis (CZE) is the simplest of a family of related analytical techniques that employ narrow-bore silica capillaries (< 100 μm i.d.) to perform high efficiency separations. CZE possesses relatively simple components—a high-voltage power supply, two electrolyte reservoirs, a capillary and commonly a UV–visible spectrophotometric detector. Cations, neutral molecules and anions can be separated in a single separation. Species of the same charge are separated depending on their mass.

Micellar electrophoretic capillary chromatography (MECC) employs a different separation mechanism from CZE and has been widely applied in recent years to the separation of non-ionic compounds. A surfactant, such as sodium dodecyl sulphate (SDS), is added to the background electrolyte at supramicellar concentrations and analytes, during migration down the capillary, partition into the lipophilic core to greater or lesser degrees depending on their chemical structure. The resulting separation can thus be significantly different from that obtained by CZE or other electrophoretic modes of separation. This particular section[117] illustrates the use of MECC with various surfactant concentrations in the separation of water-soluble vitamins B_1 (thiamine hydrochloride), B_2 (riboflavin), B_3 (nicotinamide/niacinamide) and B_6 (pyridoxine hydrochloride).

5.3.3 Procedure

A vitamin B complex tablet (5 mg vitamin B_1, 2 mg vitamin B_2, 20 mg vitamin B_3 and 2 mg vitamin B_6, together with excipients lactose, talc, magnesium stearate, iron oxide in a tablet of total mass 120 mg) is crushed in a beaker to which 20 cm^3 of methanol–water (50 + 50 v/v) containing 20 mmol dm^{-3} sodium tetraborate are added. After mixing for 2 min, 5 cm^3 of the supernatant are withdrawn and filtered through a 0.45-μm filter into a CE sample vial (sample solution A). A second tablet is similarly crushed and mixed with 20 cm^3 of methanol–water (50 + 50 v/v) containing 25 mmol dm^{-3} SDS, 20 mmol dm^{-3} sodium tetraborate and saturated with Sudan III {1-[p-(phenylazo)phenyl]azo(2-naphthol)} (**XXXVI**) as the micelle marker.

(**XXXVI**)

After mixing for 2 min, 5 cm^3 of the supernatant are withdrawn and filtered as above into a CE sample vial (sample solution B). These samples are then injected (injection volume 4 nl) into a 50 μm i.d. silica capillary with a voltage gradient of 40 kV m^{-1} operating at a temperature of 30 °C. The detection wavelength is 200 nm and the background electrolyte is sodium tetraborate–sodium dihydrogenphosphate (20 mmol dm^{-3}, pH 9) containing various concentrations of SDS (0, 50, 100 and 150 mmol dm^{-3}). Typical data for these experiments are given in Table 32.

The order of elution when [SDS] = 0, i.e. using the CZE mode, correlates with the net charge on the vitamins at the pH of the background electrolyte, i.e. pH 9, with cations being eluted before neutral molecules and the latter before anions. Vitamin B_1, which has a single positive charge (**XXXVII**), is eluted first, followed by the neutral vitamin B_3 (**XXXVIII**), which co-elutes with the neutral marker (methanol). Vitamin B_2 (**XXXIX**) has a pK_a value corresponding to deprotonation of the amide

Table 32 Variation of migration times (min) for selected vitamins with increasing SDS concentration

[SDS] (mmol dm^{-3})	t_{B_1}	$t_0{}^a$	t_{B_3}	t_{B_2}	t_{B_6}	$t_{mc}{}^b$
0	3.41	4.69	4.69	6.31	8.31	—
50	6.54	3.99	4.36	5.49	5.93	12.55
100	8.43	4.12	4.92	7.00	6.16	14.9
150	11.21	4.14	5.58	8.81	6.59	17.7

^{a} t_0 = Migration time of the neutral marker.
^{b} t_{mc} = Migration time of the micelle marker.

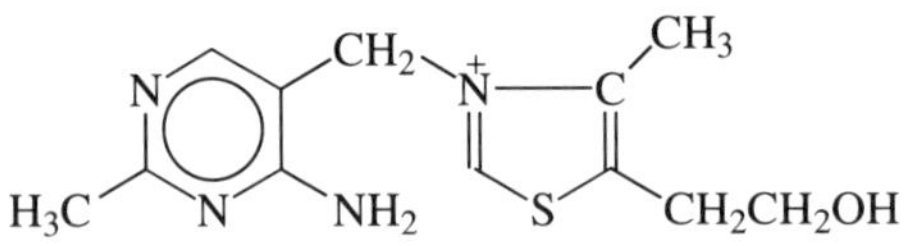

(XXXVII)

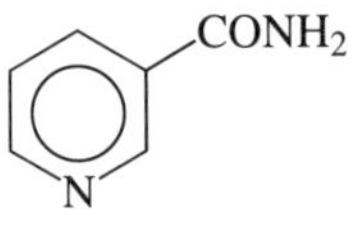

(XXXVIII)

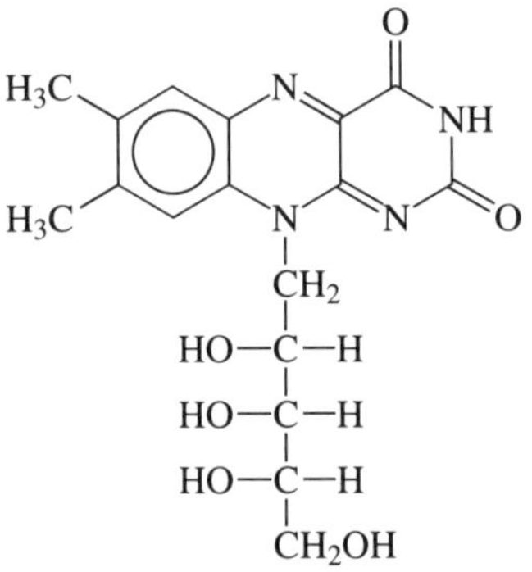

(XXXIX)

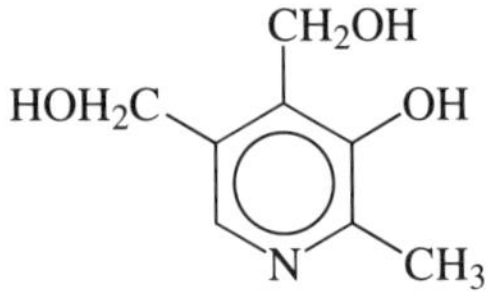

(XL)

nitrogen at 9.7, so at pH 9 it possesses a partial negative charge (degree of dissociation $\alpha \approx 0.2$). It elutes next, followed by vitamin B_6 (**XL**), which has a pK_a value of 9, corresponding to deprotonation of the phenolic entity (degree of dissociation $\alpha = 0.5$). The degree of dissociation, α, is calculated from an expression of the Henderson–Hasselbach equation:

$$pH = pK_a + \log\left(\frac{\alpha}{1-\alpha}\right) \tag{36}$$

The effect of increasing [SDS] is most dramatically reflected in the change in the migration time of vitamin B_1, suggesting very strong binding within the micelles so that it now, even with its positive charge, elutes after vitamins B_2, B_3 and B_6 but before the neutral micelle marker. Vitamin B_2 shows moderate binding whereas vitamin B_3 shows only a slight increase in migration time with increasing [SDS], indicating very little binding, if any. Vitamin B_6 shows anomalous behaviour with its migration time slightly decreasing with increasing [SDS].

Following identification of the peaks due to vitamins B_1, B_2, B_3 and B_6, their concentration in the injected solutions A and B can be determined by reference to calibration plots constructed for each vitamin or by resort to the internal standard method of quantitation using, for example, o-ethoxybenzamide (Section 2.8.1). The concentration in mg vitamin per tablet can then be calculated for each vitamin. This simultaneous assay of these vitamins is relatively rapid and simpler than HPLC methods, which require ion pairing or gradient elution for the successful assay of such multivitamin formulations.

5.4 AIR-SEGMENTED CONTINUOUS-FLOW VISIBLE SPECTROPHOTOMETRIC DETERMINATION OF CEPHALOSPORINS IN DRUG FORMULATIONS BY ALKALINE DEGRADATION TO HYDROGEN SULPHIDE AND FORMATION OF METHYLENE BLUE[118]

5.4.1 Summary

Cephalosporins (**XLI**) can be determined by an automated method based on alkaline degradation to H_2S, formation of methylene blue followed by its determination at 667 nm using an on-line visible spectrophotometric flow cell. The sample time is 36 s so the method is ideally applied to repetitive analysis in the pharmaceutical industry. The method can also be used for the determination of trace cephalosporins and other S^{2-}-producing impurities in penicillin G and V samples since it possesses LODs in the range 0.3–0.9 $\mu g\ cm^{-3}$. It is, however, not stability indicating, since several cephalosporin degradation products are known to degrade to S^{2-} under the experimental conditions.

5.4.2 Introduction

A manual visible spectrophotometric method for the determination of cephalosporins (**XLI**) based on alkaline hydrolysis to H_2S and formation of methylene blue (Figure 38) has been adapted for use with an air-segmented AutoAnalyser I system for the determination of 12 cephalosporins in drug formulations and for determining trace amounts of cephalosporins and other sulphide-producing impurities in penicillin G and V samples (Figure 39).

5.4.3 Procedure

Figure 40 shows the signals obtained for different concentrations of cephalexin under optimum reaction conditions. Linear calibration is found in the range 8–80 $\mu g\ cm^{-3}$ for 12 cephalosporins and detection limits (twice the standard deviation of the blank) are in the range 0.3–0.9 $\mu g\ cm^{-3}$.

This automated method can be used for the determination of trace cephalosporins and other S^{2-}-producing impurities in penicillin G and V samples. It is, however, not stability indicating, since several cephalosporin degradation products are known to degrade further to sulphide. In this respect, the method is inferior to the imidazole method for cephalosporins and penicillins,[119,120] which is stability indicating owing to the direct reaction of the imidazole with the β-lactam ring (Figure 41).

5.5 USE OF IR SPECTROMETRY TO IDENTIFY AN ACTIVE RAW MATERIAL PROCAINE PENICILLIN G USED IN THE PHARMACEUTICAL INDUSTRY

5.5.1 Summary

Infrared spectrometry is a powerful tool for the identification of active raw materials used in the pharmaceutical industry. The antibacterial procaine penicillin (**XLII**) is

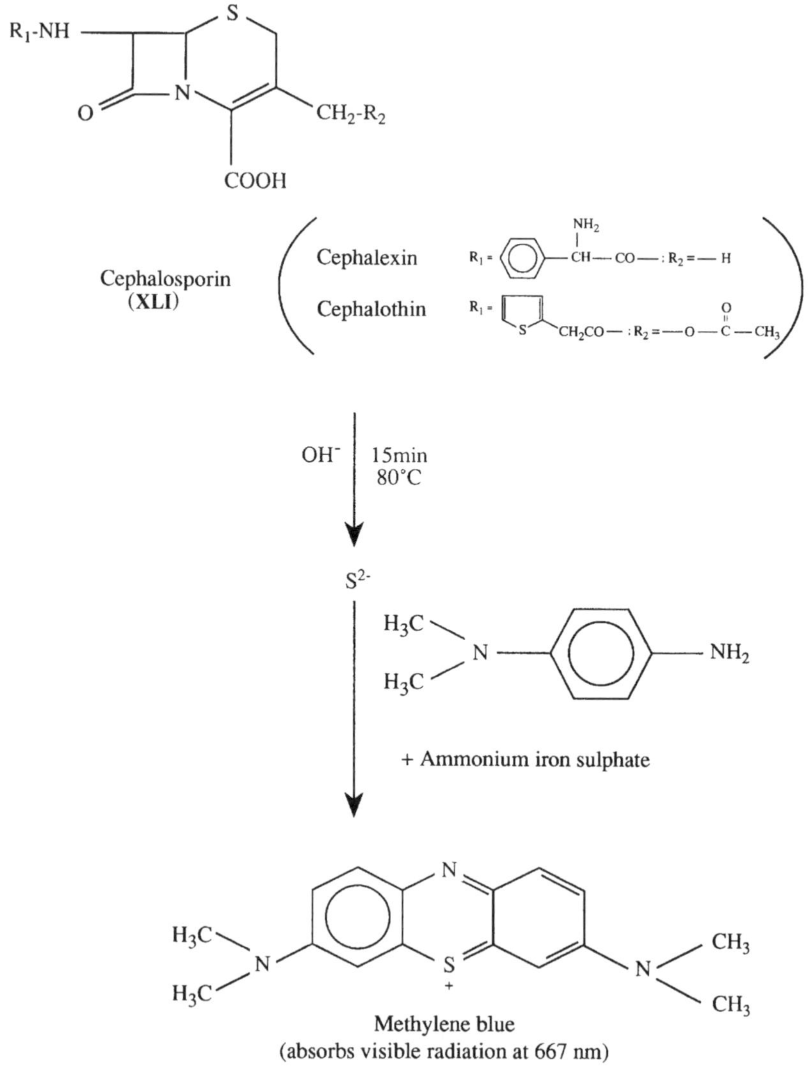

Figure 38 Scheme illustrating alkaline hydrolysis to H_2S followed by formation of methylene blue

chosen as the example and various functional groups in the overall molecule are also identified. Fourier transform IR spectrometry is significantly faster, with a full-range spectrum being recorded in about 1 s. Identification of the molecule in this case is by recourse to an IR digital spectral library.

5.5.2 Introduction

Procaine Penicillin Injection BP is a salt of penicillin G and is given intramuscularly, liberating the antibacterial agent benzylpenicilllin over 12–24 h according to the

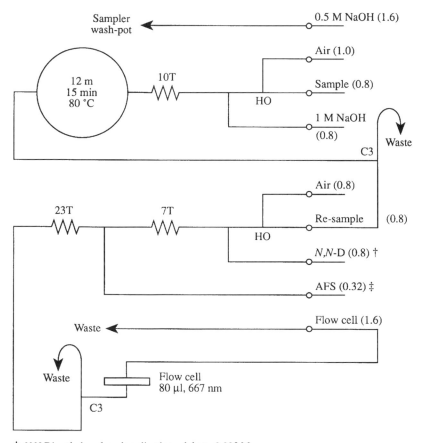

† N,N-Dimethyl-p-phenylenediamine sulphate, 0.005 M
‡ Ammonium iron(III) sulphate, 0.25 M

Figure 39 Recommended AutoAnalyzer manifold. Sample time, 36 s; wash time, 72 s. Numbers in parentheses are the flow rates in cm^3 min^{-1}. Reproduced by permission of the Royal Society of Chemistry from A. Abdalla and A. G. Fogg, *Analyst*, 1983, **108**, 53

dose administered. It is used in both human and veterinary medicine. Procaine Penicillin BP as a raw material, must conform to BP specifications regarding description, solubility, identification, acidity/alkalinity, water content, assay and other parameters such as pyrogens. The British Pharmacopaeia 1980 states that the drug must conform to three identification tests, namely (a) that the IR absorption spectrum is concordant with the reference spectrum of procaine penicillin, (b) that penicillinase causes an orange solution of the indicator neutral red, in which is dissolved procaine penicillin, rapidly to turn red and (c) that the drug yields the characteristic reaction of primary aromatic amines since procaine itself contains this functional group. An orange–red precipitate of an azo dye is therefore produced.

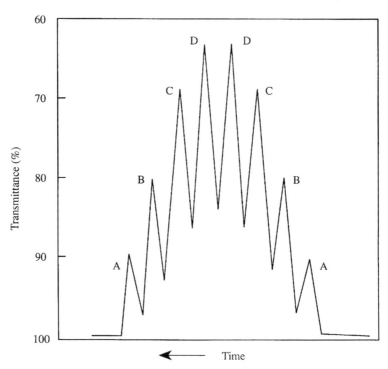

Figure 40 Signals obtained for calibration with standard cephalexin solutions. Cephalexin concentrations of solutions: A, 8; B, 16; C, 24; D, 32 µg cm^{-3}. Reproduced by permission of the Royal Society of Chemistry from A. Abdalla and A. G. Fogg, *Analyst*, 1983 **108**, 53

5.5.3 Procedure

The IR spectrum of procaine penicillin (**XLII**) should be recorded using either a conventional dispersive instrument or a non-dispersive Fourier transfrom IR spectrometer. The IR spectrum recorded on a conventional dispersive instrument is given in Figure 42, together with some functional group designations. This can be used to identify procaine penicillin as a raw material by matching this spectrum with a reference spectrum.

(**XLII**)

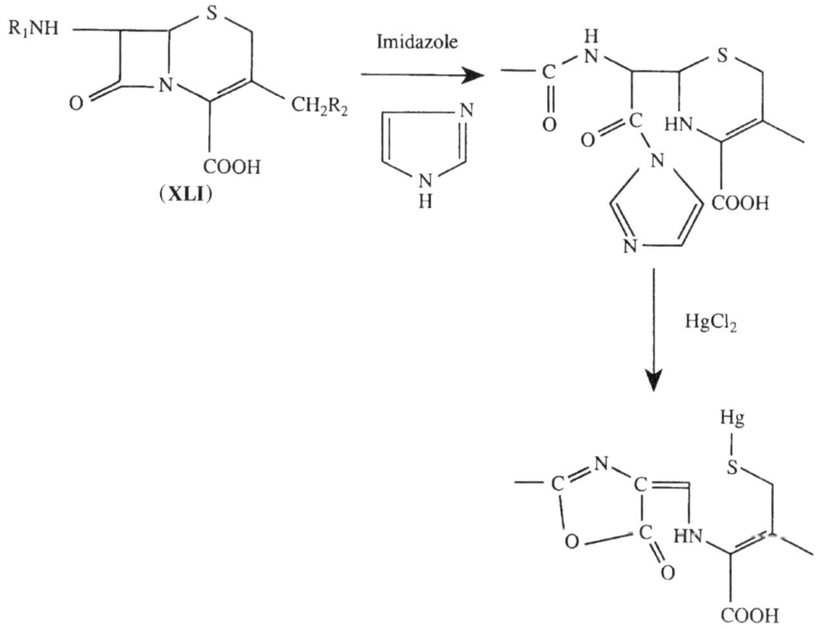

Organic Analytes Which are Major or Minor Sample Constituents 161

Figure 41 Scheme illustrating imidazole method for cephalosporin determination

Infrared spectrometers employing an interferometer and having no monochromator are now predominant. They have increased sensitivity and can record spectra more rapidly. This is because instead of scanning a spectrum over a given wavenumber range, which can take 1–4 min, the interferometer enables all the data to be collected virtually simultaneously in the form of an interferogram, then mathematically transformed (using Fourier integrals) by computer into a conventional spectrum. A full-range IR spectrum can be recorded in about 1 s and the sensitivity can be enhanced by accumulating multiple scans which are then computer processed to increase the signal-to-noise ratio. Following the recording of the IR spectrum by the FT-IR spectrometer, various computer manipulations, such as the use of IR digital spectra libraries, can be used to help identify the unknown organic molecule. This is illustrated in Figure 43.

5.6 QUALITATIVE AND QUANTITATIVE ANALYSIS OF A POLYMERIC MATERIAL USING THERMOGRAVIMETRIC ANALYSIS (TGA)[42]

5.6.1 Summary

Thermogravimetric analysis can be used to characterise a complex solid polymeric material. Quantitative analysis, with a coefficient of variation of at best 1%, is possible by taking a step in the thermogram which relates to one component of the mixture.

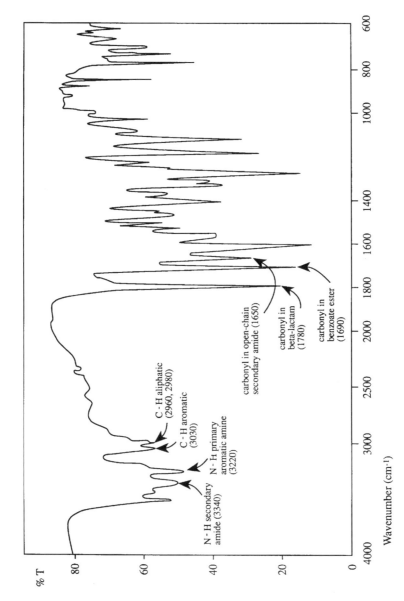

Figure 42 IR spectrum of procaine penicillin recorded on a conventional dispersive instrument

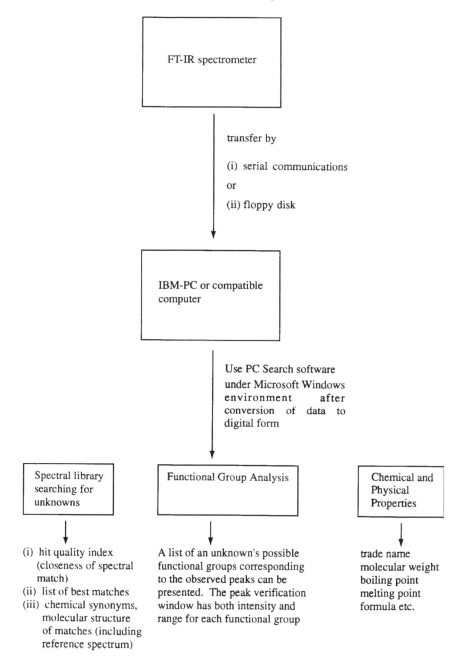

Figure 43 Combined use of FT-IR spectrometer and associated software for identification of unknown organic molecules

5.6.2 Introduction

Thermogravimetric analysis (TGA) is one of a group of thermal methods of analysis which includes differential thermal analysis (DTA; see Section 3.7), differential scanning calorimetry (DSC) and pyrolysis–gas chromatography. All of these techniques involve the study of the effect of heat on a particular sample and, in the case of TGA, the change in mass of a sample with increase in temperature is followed by a sensitive balance inside an appropriate furnace to yield a thermogram. Qualitative and quantitative analysis is possible for a wide range of sample types, e.g. (i) inorganic materials such as calcium oxalate which thermally degrades to $CaCO_3$ by loss of CO at 400–500 °C followed by a further step at 700–800 °C corresponding to loss of CO_2 to give calcium oxide, and (ii) polymeric materials such as shoe-heel rubber, which is the subject of this section. TGA is limited to samples which undergo weight changes, hence melting, crystal phase changes, etc., cannot be studied. The relative standard deviation is at best ca 1%, but is very variable.

5.6.3 Procedure

TGA can be used in the case of the polymeric material shoe-heel rubber to characterise the material by its thermogram (Figure 44). Where the thermal processes taking place are known and where a step in the thermogram of a mixture may be clearly related to one component (e.g. decomposition of the polymer itself at 400 °C), then quantitative analysis is possible.

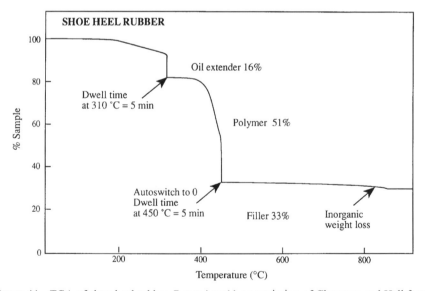

Figure 44 TGA of shoe-heel rubber. Reproduced by permission of Chapman and Hall from F. W. Fifield and D. Kealey, *Principles and Practice of Analytical Chemistry*, 3rd edn, Blackie, Glasgow, 1990

Chapter 6

Organic Trace Analysis of Low Molecular Weight Analytes in Environmental Samples and Biological Materials

6.1 DETERMINATION OF AROMATIC AND ALIPHATIC ISOCYANATES IN WORKPLACE ATMOSPHERES BY HPLC

6.1.1 Summary

Organic isocyanates can be determined in workplace atmospheres by bubbling them into a derivatising solution [e.g. 1-(2-methoxyphenyl)piperazine (MPP)] followed by reversed-phase HPLC with electrochemical detection and UV detection in series. The LOD is 0.2 μg m^{-3} based on a 10-dm^3 sample and a 10-μl injection volume and the ratio of detection responses can be used to provide additional qualitative information about the chemical nature of the particular isocyanate when retention data are deemed inadequate.

6.1.2 Introduction

The properties of isocyanates to form polymers (urethanes) are widely exploited in industry, particularly in the manufacture of upholstery products, paint, varnishes, printing inks and adhesives. Organic isocyanates are respiratory irritants and asthma-producing sensitisers and consequently a large number of industrial personnel are potentially at risk to occupational isocyanate exposure.

6.1.3 Procedure[121]

Isocyanates are collected in 10 cm^3 of a suitable derivatising solution contained in an impinger/bubbler in order to stabilise them prior to HPLC. Examples of reagents used for such derivatisations are given in Table 33. It is often advantageous, under these circumstances, to introduce an active group (chromophore, fluorophore or electrophore) into a particular analyte to improve detectability. Derivatising reagents are dissolved in a high-boiling solvent (usually toluene) to avoid excessive evaporation during the air sampling. After sampling, a 2-cm^3 aliquot of the derivatising solution is evaporated to dryness in a microreaction vessel. The residue is then taken up in 200–500 µl of a solvent compatible with the mobile phase to be used in HPLC.

A combination of UV and electrochemical detection (ED) in series can be used to provide additional qualitative information about a particular isocyanate when retention data alone may not be adequate, i.e. the ratio of detection responses of MPP-derivatised isocyanates from UV and ED will be different, for say, toluene diisocyanate (**XLIII**) and phenyl isocyanate (Figure 45).

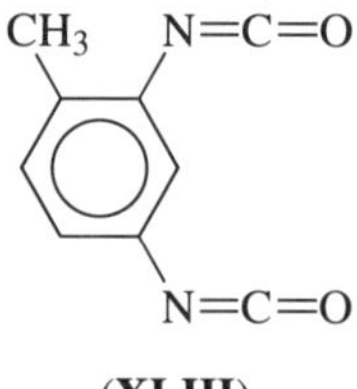

(**XLIII**)

Wu et al.[122] have determined airborne isocyanates using tryptamine (**XLIV**) as the derivatising agent followed by HPLC with dual fluorescence and amperometric detection. The characteristics of fluorescence emission and amperometric oxidation

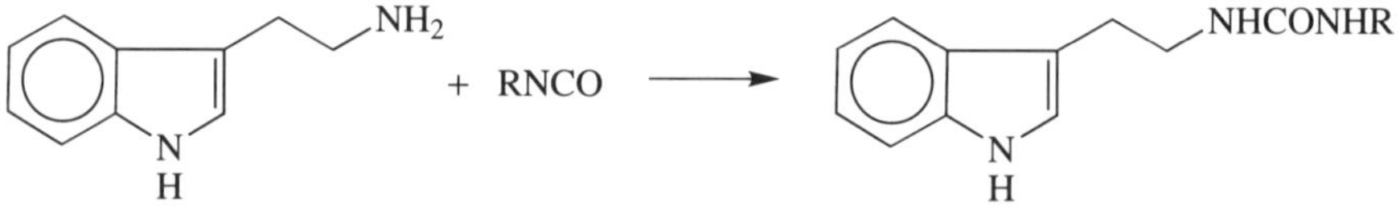

(**XLIV**)

of tryptamine are retained even after its reaction with isocyanates. All tryptamine derivatised isocyanates are capable of being calibrated with a solution of tryptamine in acetonitrile, which indicates that any of the purified derivatives is suitable for use as a reference standard, e.g. tryptamine-derivatised hexamethylene diisocyanate is

Table 33 Derivatising reagents used for isocyanate determination by HPLC

Reagent	Detection	Detection limits[a] ($\mu g \ m^{-3}$)
N-(4-Nitrobenzyl)proplyamine	UV	5
1-(2-Pyridyl)piperazine	UV	3
1-(2-Methoxyphenyl)piperazine (MPP)	UV–ED[b]	0.2

[a] Based on a 10-dm^3 sample and a 10-µl injection volume.
[b] ED = electrochemical detection at +0.8 V (vs Ag/AgCl).

relatively easy to prepare in its pure form. Since fluorescence intensity is conserved by the reacted tryptamine in the isocyanate derivatives, this clearly illustrates that the configuration of the indolyl π orbitals is not altered in the reaction. Both fluorescence and amperometric detectors (oxidation of indolyl group) are highly selective with low LODs, hence the likelihood of interferents being detected simultaneously by both detectors is minimised.

6.1.4 Calculation

Air containing isocyanates is sampled in a factory manufacturing paints and varnishes. A 10-dm^3 volume of air is sampled in an impinger/bubbler containing 10 cm^3 of the derivatising agent 1-(2-methoxyphenyl)piperazine dissolved in toluene. A 2-cm^3

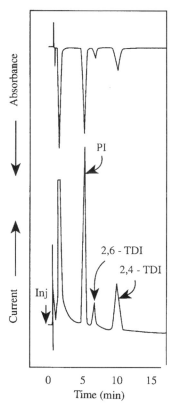

Figure 45 Chromatogram of the phenyl isocyanate (PI) and toluene diisocyanate (TDI) derivatives of 1-(2-methoxyphenyl)piperazine using both UV detection at 242 nm and electrochemical oxidation at +0.80 V (vs Ag/AgCl). A Hypersil ODS column (150 mm × 4.5 mm i.d.) was used with 40% acetonitrile–acetate buffer (pH 6) with a flow rate of 1.5 cm^3 min^{-1}. Reproduced with permission from C. J. Purnell, D. A. Bagon and C. J. Warwick, in *Analytical Techniques in Environmental Chemistry*, ed. J. Albaiges, Pergamon Press, 1982, p. 209. Crown Copyright

aliquot of this solution is then sampled, evaporated to dryness and the residue taken up in 200 μl of mobile phase, acetate buffer (pH 6)–acetonitrile (60 + 40), prior to a 10-μl injection into the HPLC system with a 15-cm C_{18} column. Using a flow rate of 0.5 cm^3 min^{-1} and electrochemical detection at +0.8 V, substances A and B elute at 16.40 and 17.63 min with peak widths of 1.11 and 1.21 min and peak areas of 240 and 120 arbitrary units, respectively. Quantities of 10 μl of 1 ppm stock solutions of 2,6-toluene diisocyanate and 2,4-toluene diisocyanate in toluene are then diluted to 10 cm^3 with the derivatising agent dissolved in toluene and allowed to react as above for the factory air sample. Following an identical procedure as for the factory air sample, these two derivatised isocyanates were found to elute at retention times which identified substances A and B as 2,6-toluene diisocyanate and 2,4-toluene diisocyanate, respectively. Equal peak areas of 20 arbitrary units were observed. Calculate (a) the column resolution, (b) the efficiency of the column in terms of number of the average number of plates, (c) the average plate height, (d) the length of column required to achieve a resolution of 1.5, (e) the time required to elute 2,4-toluene diisocyanate on the longer column and (f) the concentrations in μg m^{-3} of the two isocyanates in the factory air sample.

(a) The equation for column resolution (R) has already been given in Section 2.2, i.e. equation 7:

$$R = \frac{2\Delta t_R}{W_A + W_B} = \frac{2 \times (17.63 - 16.40)}{1.11 + 1.21} = 1.06$$

(b) The efficiency of a column in terms of the number of plates is given by the equation

$$N = 16\left(\frac{t_R}{W}\right)^2 \tag{37}$$

where t_R and W are the retention time and peak width of the base, respectively (in the same units). Therefore,

$$N_A = 16\left(\frac{16.40}{1.11}\right)^2 = 3493$$

and

$$N_B = 16\left(\frac{17.63}{1.21}\right)^2 = 3397$$

Hence

$$N_{average} = \frac{N_A + N_B}{2} = 3445$$

(c) The average plate height can be calculated from

$$H = \frac{L}{N} \tag{38}$$

where L = length of column and N = average number of plates. Therefore,

$$H = \frac{L}{N} = \frac{15}{3445} = 4.4 \times 10^{-3} \text{ cm}$$

(d) The resolution R is related to efficiency/number of plates by the equation

$$R = \frac{\sqrt{N}}{4}\left(\frac{\alpha-1}{\alpha}\right)\left(\frac{k'_B}{1+k'_B}\right) \tag{39}$$

where N is the average number of plates (N_B has also been used in calculations), k'_B is the capacity factor for the more strongly held substance B:

$$k'_B = \frac{(t_R)_B - t_{\text{solvent}}}{t_{\text{solvent}}} \tag{40}$$

and α is the selectivity factor:

$$\alpha = \frac{k'_B}{k'_A} \tag{41}$$

Since k'_B and α do not change with increasing N and L,

$$\frac{R_1}{R_2} = \frac{\sqrt{N_1}}{\sqrt{N_2}}$$

where subscript 1 refers to the original 15-cm column and subscript 2 to the longer column. Then,

$$\frac{1.06}{1.5} = \frac{\sqrt{3445}}{\sqrt{N_2}}$$

and therefore

$$N_2 = 3445\left(\frac{1.5}{1.06}\right)^2 = 6.9 \times 10^3$$

Since

$$L = N \times H \text{(equation 38)}$$
$$L = 6.9 \times 10^3 \times 4.4 \times 10^{-3} \text{ cm}$$
$$= 30.36 \text{ cm}$$

Hence to achieve an ideal baseline resolution of $R = 1.5$, the length of the column would have to be effectively doubled.

(e) The retention time of the more strongly held substance B, $(t_R)_B$, is given by the equation

$$(t_R)_B = \frac{16R^2 H}{v}\left(\frac{\alpha}{\alpha-1}\right)^2 \frac{(1+k'_B)^3}{(k'_B)^2} \tag{42}$$

where v is the mobile phone velocity and R, H, α and k'_B are as given above. Thus,

$$\frac{(t_R)_B \text{ on column 1}}{(t_R)_B \text{ on column 2}} = \frac{(R_1)^2}{(R_2)^2}$$

$$\frac{17.63}{(t_R)_B \text{ on column 2}} = \frac{(1.06)^2}{(1.5)^2}$$

$$(t_R)_B \text{ on column 2} = 35 \text{ min}$$

Hence to achieve an ideal baseline resolution of $R = 1.5$ the time of separation is also effectively doubled.

(f) Here, it is assumed that there is a linear relationship between peak area and concentration of derivatised isocyanate in the concentration range in which both the factory air sample and the standards are found. For 2,6-toluene diisocyanate and 2,4-toluene diisocyanate standards, a solution containing 1 ppb (1 μg dm^{-3}) of each of these isocyanates is subjected to derivatisation. Following the procedure of taking a 2-cm^3 aliquot, evaporation to dryness, taking up the residue in 200 μl and injection into the HPLC system, the resulting peak area is 20 arbitrary units for both isocyanates.

It therefore follows that for 2,6-toluene diisocyanate in the factory air sample, 240 arbitrary units corresponds to $(240/20) \times 10^{-2}$ μg in the original 10-cm^3 derivatising solution, i.e.

$$12 \times 10^{-2} \text{ } \mu\text{g per 10 dm}^3 \text{ air} = 12 \times 10^{-3} \text{ } \mu\text{g dm}^{-3} = 12 \text{ } \mu\text{g m}^{-3}$$

Similarly for 2,4-toluene diisocyanate in the factory air sample, 120 arbitrary units corresponds to 6 μg m^{-3}.

It should be noted that the maximum exposure limit (EH 40/95, *Occupational Exposure Limits 1995*; see Section 4.6.2) for total isocyanates (as —NCO) is given as 20 μg m^{-3} as an 8-h TWA or 70 μg m^{-3} as a 15-min TWA. The above results show that the total isocyanate concentration in the worked calculation fall within the legal control limits for these substances.

6.2 DETERMINATION OF TRIAZINE PESTICIDE RESIDUES IN ENVIRONMENTAL SAMPLES BY ENZYME-LINKED IMMUNOSORBENT ASSAY (ELISA)

6.2.1 Summary

Triazine pesticide residues can be measured simply, rapidly and inexpensively by enzyme-linked immunosorbent assay (ELISA) kits, both on-site and in the laboratory. The triazines in the sample compete with enzyme-labelled triazine for antibody molecules coated on the wall of the vessel. Following rinsing of the vessel, the enzyme label converts added substrate to product which reacts with a chromogen, changing it from colourless to blue. The reaction is stopped by the addition of acid which turns the chromogen colour to yellow, which is monitored at 450 nm. High absorbance is related to a low concentration of triazines in the original sample. The test is selective for triazines and compares favourably with HPLC and GC–MS.

6.2.2 Introduction

The presence of pesticide residues in environmental samples such as water and soil is an escalating problem that has aroused public concern over potential health hazards. As a result, the legislation relating to pesticides has become more and more stringent over the years, requiring larger numbers of more sensitive tests to be performed on such samples. In 1985, the EEC issued a Drinking Water Directive (80/778/EEC) setting standards for the quality of water for human consumption, irrespective of its source. In this Directive, pesticides are considered as a single group, with a Maximum Allowable Concentration (MAC) set at 0.1 μg dm^{-3} for any individual substance in the group and 0.5 µg dm^{-3} for total pesticides and related products. It is also important to determine pesticide concentrations for economic reasons. For instance, if a maize crop has been triazine protected, the residual triazine concentration in the soil should be monitored prior to using the same field to grow another vegetable which is affected by triazine, e.g. crops such as alfalfa and soy beans are regarded as atrazine sensitive and it is believed that atrazine can persist in soil for one or more years after application.[123]

Analytical techniques such as HPLC and GC–MS allow for the accurate determination of pesticide residues. However, they require sophisticated and expensive equipment handled by skilled technicians. There is therefore a need for simple, fast and inexpensive tests allowing technicians to perform easily large numbers of screening tests in the field or in the laboratory for a variety of pesticide residues such as triazines. ELISA kits such as Millipore EnviroGard test kits are designed to answer such needs. Such an ELISA kit contains plastic test tubes precoated with the relevant antibody and the reagent solutions needed to perform the test.

6.2.3 Procedure

(a) An environmental sample such as a water sample containing the triazines (soil samples can be extracted with acetonitrile prior to dilution of a small amount of the extract with laboratory grade water[124]) or calibration solutions of the triazine are added to test-tubes whose walls are coated with antibodies that are specifically directed against the triazine to be detected. This accounts for the selectivity of the test. It should be noted that there is the same quantity of antibody molecules in each test-tube.
(b) The first reagent is added, i.e. a solution of pesticide covalently bound to a chosen enzyme, otherwise known as the enzyme conjugate. The free pesticide introduced in (a) and the enzyme conjugate compete to bind to the antibody molecules coated on the tube wall.
(c) After incubation/equilibration, the tubes are washed several times with water. Only the unlabelled and labelled pesticide molecules which bind to the antibody will remain inside the tube. The quantity of enzyme labelled pesticide bound will be inversely proportional to the quantity of pesticide present in the sample or standard solution added in (a).
(d) A solution of substrate (S) is added. Then, immediately afterwards, a solution of chromogen (C) is added.
(e) The enzyme will catalyse the transformation of substrate (S) into product (P), which in turn will react with the chromogen (C) to change it from colourless to blue. Each enzyme molecule will catalyse the conversion of thousands of substrate molecules into product molecules reacting with the chromogen. This multiplication effect accounts for the low LOD of the test.
(f) After a short reaction time, a 'stop solution' (dilute acid) is added into the test-tube. The effect of this solution is immediately to stop the reaction by destroying

the enzyme. It also changes the dye colour from blue to yellow. The absorbance of the solution is then measured at 450 nm. A high concentration of pesticide in the sample or standard solution will result in a lighter yellow colour.

EnviroGard test kits have been successfully used for fast on-site and laboratory screening to monitor pesticides such as triazines in environmental samples.[125] The kits provide all the reagents needed to perform the semi-quantitative determination of pesticide in only 7 min. The antibody's high selectivity is illustrated in Table 34, which shows the concentration in μg dm^{-3} of selected triazines needed to reduce by 50% the conversion of substrate to product in the EnviroGard Triazine tube kit.[126]

The binding is noticeably stronger for those molecules with similar structures to atrazine (**XLV**), i.e. having 4-ethylamino and 6-isopropylamino groups, e.g. ametryn (**XLVI**), but noticeably weaker for those molecules that lack this configuration, e.g. cyanazine (**XLVII**).

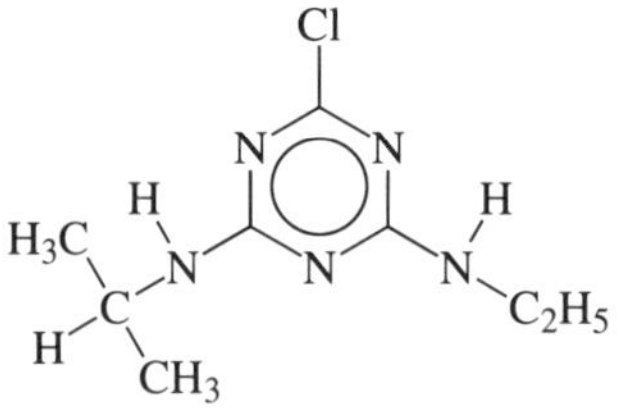

(XLV)

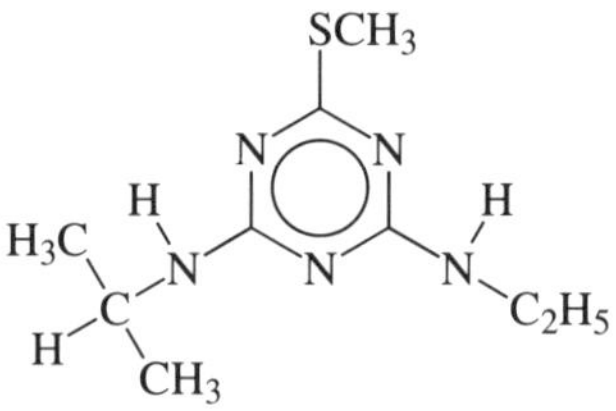

(XLVI)

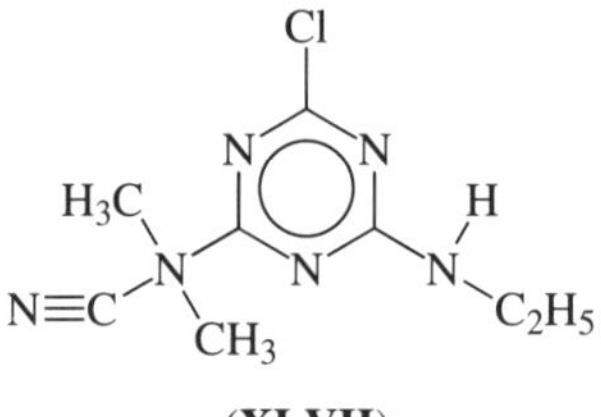

(XLVII)

Table 34 Concentration of selected triazines needed to reduce by 50% the conversion of substrate to product in the EnviroGard Triazine tube kit

Triazine	Concentration (μg dm^{-3})
Atrazine	0.40
Ametryn	0.45
Prometryn	0.50
Propazine	0.50
Hydroxyatrazine	28.00
Cyanazine	40.00
Didealkylatrazine	No response

LMW Organics in Environmental Samples and Biological Materials 173

Table 35. Comparison of EnviroGard kit with HPLC for the determination of atrazine in water.[127]

Sample water source	Atrazine added (ppb)	Determined by EnviroGard kit (ppb)	Determined by HPLC (ppb)
River	10	11	12
Well	4	3	4
Blueberry waste water	10	9	8
Pond	10	10	10

EnviroGard kits such as those for triazines give determinations that correlate reasonably well with alternative analytical techniques such as HPLC (Table 35) and GC–MS. A plot of immunoassay determinations in $\mu g\ dm^{-3}$ for 66 surface water samples in central USA against the GC–MS summed response in $\mu g\ dm^{-3}$ gave linear regression analysis with $r = 0.95$, slope 0.84 and intercept 0.0058 over the concentration range 0.05–3 $\mu g\ dm^{-3}$.[126]

6.3 ANALYSIS OF ALCOHOLIC BEVERAGES

6.3.1 Summary

Ethyl acetate and higher alcohols can be identified and determined in distilled liquors by direct injection on to a gas chromatographic column with flame ionisation detection using the internal standard method of quantitation. The ratio of the concentration of 2-methylbutanol to that of 3-methylbutanol can be used in certain cases to characterise a good-quality spirit.

6.3.2 Introduction

According to the AOAC Official Methods of Analysis (1984), beverages such as beers, wines and spirits are assayed for organic chemical entities such as ethanol and aldehydes by both classical and instrumental methods of analysis. Classical determinations are generally time consuming, give concentrations in terms of the major impurity whether it be acid, ester, higher alcohol, etc., and are relatively insensitive, especially to trace constituents. Instrumental methods such as GLC and HPLC are therefore increasingly applied to the identification and determination of organic impurities in beverages. Atomic absorption spectrometry has found widespread acceptance for the identification and determination of metal ions such has Ca^{2+}, Cu^{2+} and Fe^{3+} in beverages.

Ethanol itself is determined by first neutralising the sample to prevent interference from volatile acids such as acetic acid and is then distilled. The distillate is made up to the original volume of sample and the specific gravity and/or refractive index is measured. The percentage of ethanol by volume at a particular temperature can then be read off from tables (Table 36) when using an immersion refractometer.

Table 36 Determination of percentage of ethanol by volume using an immersion refractometer

Refractive index	Temperature (°C)		
	17.5	20	23
1.33319	0.00	0.52	1.16
1.33397	1.63	2.11	2.69
1.33513	3.83	4.32	4.90
1.33590	5.27	5.76	6.34

Esters can be determined by classical methods by taking the distillate from the ethanol determination, neutralising it and reacting it with NaOH under reflux to saponify the esters to the corresponding carboxylic acids and alcohols. The esters are then determined as ethyl acetate following acid–base titration of the carboxylic acids. Aldehydes can be determined by addition of excess $NaHSO_3$ to the above-mentioned distillate to form hydrogen-sulphite addition compounds. After this reaction, excess I_2 is added to react with the remaining $NaHSO_3$. The remaining I_2 is then titrated with standard $Na_2S_2O_3$.

Ethyl acetate and higher alcohols (propan-1-ol, 2-methylpropanol, 3-methylbutanol) in distilled liquors can be identified and determined by the use of the internal standard butan-1-ol by GLC with flame ionisation detection using 23% Carbowax on Chromosorb W as the stationary phase. The detector and inlet temperature is maintained at 150 °C, the column temperature at 70 °C (isothermal) and the helium carrier gas flow rate at 150 cm^3 min^{-1}.[128]

6.3.3 Procedure

The standard solution is prepared by making up 1 cm^3 of propan-1-ol, 1 cm^3 of 2-methylpropanol, 2 cm^3 of 3-methylbutanol and 1 cm^3 of ethyl acetate to a total volume of 100 cm^3 with 40% ethanol. A 10-cm^3 volume of this solution is diluted to 200 cm^3 with 40% ethanol followed by dilution of 5 cm^3 of the resulting solution together with 1 cm^3 of the internal standard solution (1 cm^3 of butan-1-ol diluted to 200 cm^3 with 95% ethanol) to 100 cm^3 with 95% ethanol. The final solution contains 2.0 g of propan-1-ol, 2.1 g of 2-methylpropanol, 4.1 g of 3-methylbutanol and 2.3 g of ethyl acetate per 100 dm^3, together with the internal standard (4.1 g of butan-1-ol per 100 dm^3).

Identification and determination of ethyl acetate and the higher alcohols are then carried out as follows:

(a) Preliminary injections of 10-μl amounts of the sample made to determine the absence of the internal standard, butan-1-ol.
(b) A 1-cm^3 volume of the internal standard solution is added to 100 cm^3 of the sample and 10-μl aliquots are chromatographed in triplicate. This results in the sample chromatogram.

(c) Aliquots of 10 μl of the standard solution containing 2.1 g of 2-methylpropanol per 100 dm^3, etc., are then chromatographed in triplicate to yield the standard chromatogram.
(d) Peak areas of ethyl acetate and the higher alcohols are then measured in both chromatograms and the peak-area ratios of each analyte to the internal standard, butan-1-ol, are calculated. It is then possible using this internal standard approach to quantify the concentration of ethyl acetate and the higher alcohols in the beverage as follows:

$$\frac{\text{concentration of analyte in g/100 dm}^3 \text{ in the sample}}{\text{concentration of analyte in g/100 dm}^3 \text{ in the standard}} = \frac{\text{peak-area ratio of analyte in the sample chromatogram}}{\text{peak-area ratio of analyte in the standard chromatogram}}$$

A typical chromatogram for a Scotch all-malt whisky using the stationary phase polyethylene glycol 200 is given in Figure 46, showing the essential absence of butan-1-ol (the internal standard) and peaks due to ethyl acetate and four higher alcohols.[129] The fusel oil ratio of spirits, i.e. [2-methylbutanol]/[3-methylbutanol], can be used, in certain cases, for the characterisation of a particular spirit, e.g. for cognacs/brandies where good products (as determined organoleptically) have ratios in the region of 0.19–0.24 and inferior ones 0.24–0.26. These latter products also have relatively high concentrations of methanol and butanols.[129]

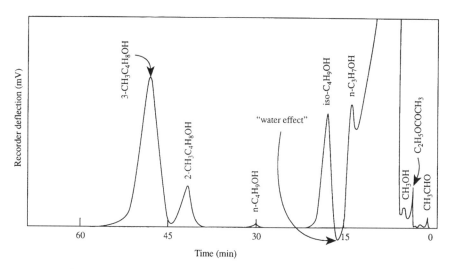

Figure 46 Gas chromatogram of a Scotch all-malt whisky. Reproduced by permission of the Royal Society of Chemistry from S. Singer, *Analyst*, 1966, **91**, 790

6.4 DETERMINATION OF CONTAMINATING ANTIBIOTIC TRACE CONCENTRATIONS IN DAIRY FEEDSTUFFS BY GC–MS[130]

6.4.1 Summary

Lincomycins A and B and clindamycin can be quantitated at trace levels in dairy feedstuffs by GC–MS following a clean-up procedure using solid-phase extraction. The major peaks in the electron impact mass spectrum were used for chromatographic detection using selected-ion monitoring. The LOD of the analytical method was 0.1 ppm.

6.4.2 Introduction

It was observed that ketosis was induced in cattle fed a new mix of feedstuffs. It was suspected that the feed mills, used to prepare the feedstuffs, had also been used for the preparation of certain pig feedstuffs known to contain the antibiotic lincomycin under veterinary prescription. It was therefore necessary to identify and determine trace levels of lincomycin in the aforementioned cattle feedstuffs by GC–MS.

Lincomycin is part of a small family of antibiotics (**XLVIII**). It was readily demonstrated that simple gas chromatographic methods would be unsuitable because of the substantial interference with many other compounds which were present in the feed extracts. These had similar retention properties and swamped the peak due to the lincomycin derivatives tried.

R_1	R_2	
C_3H_7	OH	Lincomycin A
C_2H_5	OH	Lincomycin B
C_3H_7	Cl	Clindamycin

(**XLVIII**)

6.4.3 Procedure

The electron impact (EI) and chemical ionisation (CI) mass spectra of the silyl esters of lincomycins A and B and of clindamycin are shown in Figures 47a and b. The EI mass spectra of the silyl esters are interesting in that there is severe fragmentation—the mass ion cannot be identified and the only definitive peak in the spectrum is due to the *N*-methyl-4-propylpyrrolidinium ion (MW 126) (**XLIX**) in lincomycin A and clindamycin and the equivalent for the 4-ethyl derivative of lincomycin B. The CI mass spectra, on the other hand, as expected, had considerably less fragmentation

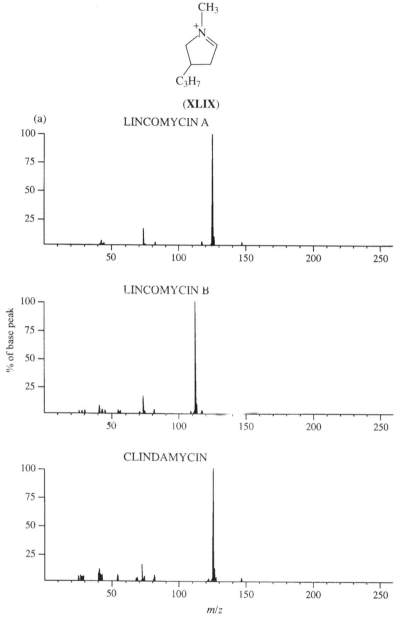

Figure 47 (a) EI mass spectra of the silyl esters of lincomycin A and B and clindamycin; base peaks occurring at m/z 126 in lincomycin A and clindamycin and at m/z 112 for lincomycin B. Reproduced by permission of the Department of Agriculture for Northern Ireland from the *Annual Report on Research and Technical Work*, Veterinary Research Laboratories, Belfast, 1982, p. 235; also published by C. McMurray et al., *J. Assoc. Off. Anal. Chem.*, 1984, **67**, 583

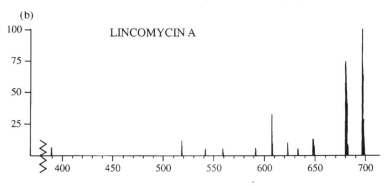

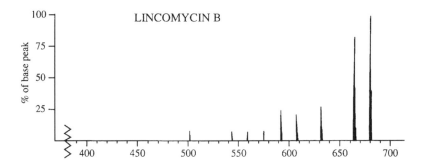

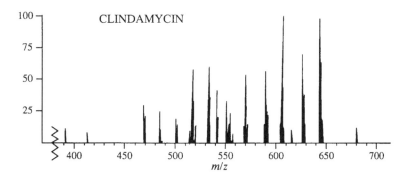

(b) Methane CI mass spectra of the tetrasilyl derivative of lincomycin A and B and of the trisilyl derivative of clindamycin. The base peak in each compound corresponds to the M + 1 ion. Reproduced by permission of the Department of Agriculture for Northern Ireland from the *Annual Report on Research and Technical Work*, Veterinary Research Laboratories, Belfast, 1982, p. 235; also published by C. McMurray et al., *J. Assoc. Off. Anal. Chem.*, 1984, **67**, 583 and the mass ions can be identified for the tetrasilyl derivatives of lincomycin A and B. The trisilyl derivative is formed from clindamycin.

The major peaks in the EI mass spectrum were used for chromatographic detection using selected-ion monitoring (SIM). In order to reduce the mass of contamination material obtained from simple methanolic extracts of feed injected on to the chromatographic column, it proved necessary to use a clean-up procedure. It was found that a simple C_{18} reversed-phase cartridge was suitable for this purpose. Conditions for adsorption and elution of lincomycin A were determined and the maximum binding of lincomycin was achieved with 20% methanol in water whereas for maximum recovery 70% methanol was suitable.

The assay developed was linear: peak area ($\times 10^{-3}$) = 3.45 (± 0.05) $\times$ lincomycin (ng) $-$ (0.93 $\pm$ 1.11) ($\pm$ values = SD). The correlation coefficient was 0.998. The LOD of the test with respect to lincomycin levels in feed was 0.1 ppm. This was equivalent to 0.5 ng of lincomycin A detected by the analytical system. The assay did not separate lincomycin A from clindamycin. Lincomycin B, on the other hand, was separated from both chromatographically and by the fact that it is identified at a different atomic mass (Figure 48). A smaller ion at m/z 73 present for all three compounds is unsatisfactory for confirming identity using ion mass ratios; there is too much interference from extraneous compounds (Figure 49).

The method developed has a considerable number of advantages over existing methods. However, the one limitation is its inability to discriminate between lincomycin A and clindamycin. In practice, this limitation is not serious as clindamycin is not generally used as a feed additive in veterinary medicine. The unique identification, however, would be readily achieved using chemical ionisation mass spectrometry as the detection technique.

Subsequent to the development of this method, lincomycin A was identified and quantitated in all feeds where the clinical and biochemical findings were consistent with the occurrence of feed-induced ketosis. It is concluded, therefore, that lincomycin A was the cause of the problem.

6.5 EXTRACTION OF COCAINE AND ITS METABOLITES FROM A URINE SAMPLE BY SOLID-PHASE EXTRACTION

6.5.1 Summary

Cocaine and its metabolites can be determined in a urine sample by gas chromatography in the concentration range 75–1000 ng cm^{-3} after a simple and relatively rapid clean-up using solid-phase extraction in mixed-mode operation. The limit of quantitation was quoted as < 10 ng cm^{-3}.

6.5.2 Introduction

The speed of analytical technological development over the last few decades is remarkable. The increasing availability of published analyses using improved inst-

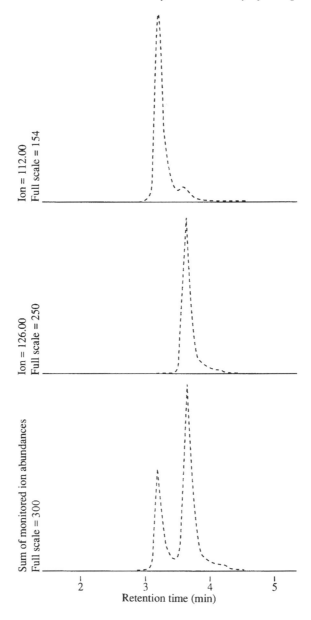

Figure 48 Separation of lincomycin B (using the ion at m/z 112) from lincomycin A (using the ion at m/z 126) and clindamycin at a ramp rate of 10°C min^{-1}. Other chromatographic conditions: injection port, 300°C; oven temperature programme, held at 120°C for 0.2 min and then ramped at 15°C^{-1} to 260°C, held for 2 min. Reproduced by permission of the Department of Agriculture for Northern Ireland from the *Annual Report on Research and Technical Work*, Veterinary Research Laboratories, Belfast, 1982, p. 235; also published by C. McMurray *et al.*, *J. Assoc. Off. Anal. Chem.*, 1984, **67**, 583

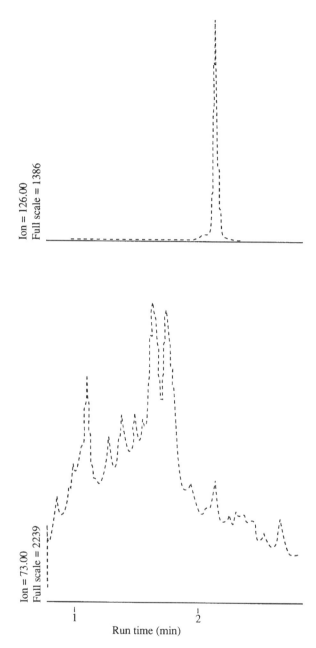

Figure 49 GC–MS profile using SIM at *m/z* 126 and 73 of a methanol extract after clean-up from a feed containing 10 ppm of lincomycin A. Reproduced by permission of the Department of Agriculture for Northern Ireland from the *Annual Report on Research and Technical Work*, Veterinary Research Laboratories, Belfast, 1982, p. 235; also published by C. McMurray *et al.*, *J. Assoc. Off. Anal. Chem.*, 1984, **67**, 583

rumentation with autosamplers and faster data handling, and LIMS for combining multiple analyses and generating instant reports, have increased productivity and cut costs. Before these analytical instruments can be used, however, samples must be prepared for analysis. This is often the hardest and most time-consuming stage of the entire process. Solid-phase extraction (SPE) makes the preparation stage faster, easier and more automatable.

Until 1978, solvent exchange, sample concentration and removal of undesired compounds from a sample required liquid–liquid extraction, during which a sample and an immiscible solvent were manually shaken and allowed to separate in a funnel. This was subject to the vagaries of an individual operator's technique, often required hazardous solvents and was not very selective. Other disadvantages of liquid–liquid extraction include the tendency of samples to form intractable emulsions when shaken up, extensive solvent use and waste disposal and the expense of glassware and distillation apparatus. Since the two liquids must be immiscible, the range of properties shown by the extracting solvent is limited by the sample solvent. For example, methanol may be an ideal solvent for an extraction, but if a particular sample is aqueous, the two liquids would mix rather than settle into separate layers.

SPE, also known as liquid–solid extraction, offers fewer disadvantages. By using a solid surface (usually based on a powdered porous silica to which organic functional groups have been bonded) as the extracting or bonded phase, the compounds that are to be extracted can be retained on the solid surface. This is equivalent to those compounds passing into the extracting solvent in a liquid–liquid extraction, but is more subtle and selective. Insolubility or solubility of the analyte with either liquid is the only property that can be used in the liquid–liquid extraction. A chemically bonded solid phase cannot mix physically to form a single phase with the liquid sample; thus SPE eliminates the need for immiscibility.

In addition, the solid support can constrain the size of molecules it can extract. If the solid support is porous, and a molecule is too large to enter the pores, it will be barely retained, compared with those that can enter the pores. Further, most organic molecules can interact with their surroundings in several different ways, such as hydrogen bonding, dipolar interactions or electrostatic (ionic) attraction. By careful control of the solid-phase bonding process, these interactions can be fine-tuned to give that surface a specific affinity for a particular range of compounds. In many ways, it brings the separating power of liquid chromatography and the speed of ordinary preparation techniques to the sample preparation step.

Solid-phase extraction cartridges are disposable cartridges that are packed with a few hundred milligrams of chemically bonded silica. Packing a sorbent in an open-topped syringe barrel creates a complete extraction unit, containing the sorbent and acting as a sample reservoir. This unit is handled singly by pushing the sample through the sorbent with positive pressure, such as that supplied by a syringe, or in batches of 10, 24 or more by employing a vacuum manifold with a port for each extraction cartridge. Various designs are now available for automated SPE–HPLC, e.g. the Varian 9200 Prospekt model.

The principles of SPE can be understood with reference to four stages:

(a) Conditioning of the sorbent prior to introduction of an aqueous sample, e.g. MeOH is used to penetrate the organic bonded layer to permit water molecules and analyte to diffuse into, rather than run over the surface of the bonded phase. Once conditioned, excess solvent is removed by washing with water in the case of an aqueous sample.
(b) A sample is applied to the conditioned column and pulled through under vacuum. Compounds that interact strongly with the sorbent are retained.
(c) In complex extractions, other liquids are used to flush out potential interferences that remain on the sorbent. The vacuum can be left on to dry the sorbent to ensure the final extract is free from unwanted water.
(d) A collection tube is placed in the vacuum manifold and a few cm^3 of an organic solvent such as ethyl acetate are applied. The solvent that passes through the sorbent is then injected directly into an analytical instrument, e.g. for GLC or HPLC.

6.5.3 Procedure[131]

A urine sample suspected of containing an abused substance such as cocaine (**L**) and its metabolite benzoylecgonine (**LI**) is passed through a preconditioned, mixed-

mode column (Figure 50). This largely aqueous sample is followed by an aqueous wash of deionized water and a solution of 0.1 mol dm^{-3} hydrochloric acid. The drug and metabolite are held on the sorbent by hydrophobic forces while strongly water-soluble or entirely inorganic species (e.g. urea, salts, proteins and sugars) are washed through the cartridge. Methanol is then applied to the cartridge. Capable of disrupting hydrophobic interactions, the methanol removes all those species still left on the column that cannot bind tightly to the ion-exchange sites, so that lipids, fatty acids, small nonpolar peptides, cholesterols and so on are flushed from the sorbent. Finally, a small volume of dichloromethane–propan-2-ol with ammonia solution is applied to elute the residual analyte. The ammonium ions displace the cocaine and benzoylecgonine cations that were held to the ion-exchange sites. The result is a simple and quick clean-up that results in a limit of quantitation of less than 10 ng cm^{-3}, and linearity with a correlation coefficient of 0.999 from 75 ng cm^{-3} (half the NIDA-required cut-off limit) to 1000 ng cm^{-3} when the measurement step is gas chromatography.

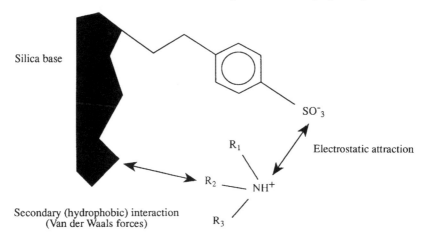

Figure 50 Mode of action of a mixed mode solid-phase extraction cartridge/column

6.6 DETERMINATION OF NITROGLYCERIN, 2,4-DINITROTOLUENE AND DIPHENYLAMINE IN GUNSHOT RESIDUE BY HIGH-PERFORMANCE LIQUID CHROMATOGRAPHY WITH ELECTROCHEMICAL DETECTION (HPLC–ED)[132]

6.6.1 Summary

High-performance liquid chromatography with oxidative and reductive electrochemical detection can be used to characterise smokeless gunpowders in terms of the content of nitroglycerin, 2,4-dinitrotoluene, diphenylamine and nitro-substituted diphenylamines. Gunshot residue can be sampled from the back of the firing hand and, after a brief preliminary treatment of sample, subjected to HPLC–ED in both oxidative and reductive modes. The resulting chromatograms are believed to be indicative of firing a weapon or handling nitroglycerin-based explosives.

6.6.2 Introduction

Detection of explosive substances has been the subject of many investigations. Law enforcement officials and forensic chemists have traditionally examined post-blast debris in search of explosive material and gunshot residue. The latter is useful in connection with suicides, and to ascertain if a suspect has fired a gun or just handled a weapon which was discharged. Detection of explosive material in post-blast debris such as that obtained from the site of the explosion, contaminated ground water and gunshot residue is difficult by gas chromatographic methods owing to the thermal instability and non-volatility of explosive compounds. Liquid chromatography is ideally suited for these determinations. A variety of detectors and detection techniques have been investigated such as UV, thermal energy analyser (TEA), off-line chemical ionisation mass spectrometry and electrochemical detection (ED). This particular problem evaluation considers the use of both reductive electrochemical detection at an amalgamated gold electrode and oxidative electrochemical detection at a glassy carbon electrode to detect and determine relevant organic molecules in gunshot residue following separation by high-performance liquid chromatography.

6.6.3 Procedure

Samples of smokeless powders were dissolved in acetone (50–300 $\mu g\ cm^{-3}$) and then diluted (10–100-fold) with the mobile phase (0.2 mol dm^{-3} H_3PO_4, 75% v/v methanol, pH 2.3) prior to injection on to a C_{18} reversed-phase HPLC column. All solutions were kept in amber-coloured glassware to prevent photodecomposition, particularly of nitrate esters such as nitroglycerin. Both injected solution and mobile phase were subjected to deoxygenation procedures. Using a 0.22 calibre Western Long Rifle Extra Power Rim Fire cartridge, the composition in g per 100 g of the electroactive components was found to be 19.0 nitroglycerin (**LII**) as determined by reductive detection, 0.32 2,4-dinitrotoluene (**LIII**) as determined by reductive detection, 0.32 diphenylamine (**LIV**) and 0.04 4-nitrodiphenylamine (**LV**) as determined by oxidative electrochemical detection.

Other smokeless gunpowders had different compositions in terms of **LII**, **LIII**, **LIV** and **LV** and also 2-nitrodiphenylamine, so that HPLC–ED could be used to characterise the particular powder.

After the weapon had been fired, the back of the firing hand was swabbed with cotton dipped in acetone. The cotton swab was placed in a brown-glass screw-capped vial to minimise photodecomposition of nitroglycerin. The swab was washed with 20% aqueous ethanol solution. After the ethanol solution had been filtered through a 0.2-μm nitrocellulose filter, a 20-μl volume of the deoxygenated solution was injected on to the HPLC column. The same procedure was used to obtain a blank sample before the weapon was fired. Both nitroglycerin and 2,4-dinitrotoluene were detected by HPLC–reductive ED at -1.00 V as well resolved peaks (Figure 51A) after this Western cartridge had been discharged and in this particular case it was found that unburned flakes of gunpowder in the gunshot residue retained their original composition of organic constituents with respect to reducible nitroglycerin and 2,4-dinitrotoluene. The determination of diphenylamine (**LIV**) in gunshot

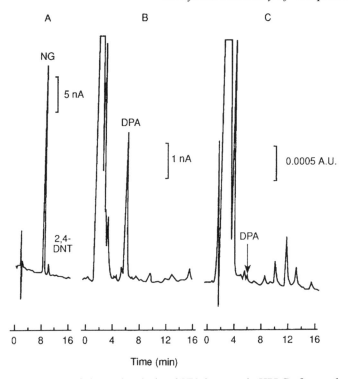

Figure 51 Comparison of electrochemical and UV detectors in HPLC of a gunshot residue. (A) reductive mode detector, Au/Hg electrode at -1.00 V, flow rate 2 cm^3 min^{-1}; (B) oxidative mode electrochemical detection at a glassy carbon electrode at ca $+1.00$ V; (C) UV detection at 254 nm. HPLC conditions: C$_{18}$ reverse-phase column, 0.2 mol dm^{-3} H$_3$PO$_4$, 75% methanol, pH 2.3 mobile phase and flow rate 1.3 cm^3 min^{-1}. Reproduced with permission of Elsevier Science from Bratin *et al.*, *Anal. Chim. Acta*, 1981, **130**, 295

residue is more difficult, since oxidative-mode chromatograms are more complex than the reductive-mode chromatograms because of interference from phenolic compounds which are present in the lipid layer on the skin, especially if handcream lotions that contain parabens have been used. Figure 51B illustrates the detection of **LIV** in gunshot residue by HPLC–oxidative ED. The LOD for diphenylamine using HPLC–oxidative ED (diphenylamine has an $E_{1/2}$ of $+0.97$ V from hydrodynamic voltammograms on a glassy carbon electrode) is still significantly superior to that with HPLC–UV, with values of 0.039 and 1.36 ng, respectively, at a signal-to-noise ratio of 3. This is illustrated by comparison of the chromatograms in Figure 51B and C.

This technique therefore provides a simple, rapid and relatively inexpensive tool for determining organic components in gunpowders. No easily reducible molecules were detected in the gunshot residue except gunpowder components. Detection of nitroglycerin and other components of smokeless powders is therefore indicative of firing a weapon or handling nitroglycerin-based explosives.

6.7 DETERMINATION OF PROZAC (FLUOXETINE) AND ITS DEMETHYLATED METABOLITE IN SERUM BY HPLC WITH FLUORESCENCE DETECTION[133]

6.7.1 Summary

The designer drug Prozac can be determined in serum by an analytical method which involves (i) solid-phase extraction (SPE) to retain Prozac and its N-demethylated metabolite and to separate them from interfering substances in the complex matrix and (ii) elution of **LVI** and **LVII** and HPLC with fluorimetric detection. The method is more specific than HPLC–UV and can detect 20 ng cm^{-3} of either fluoxetine using 0.5 cm^3 of sample.

6.7.2 Introduction

Prozac (fluoxetine) (**LVI**) is a bicyclic antidepressant which enhances serotoninergic neurotransmission through potent and selective inhibition of neuronal reuptake of serotonin. Its N-demethylated metabolite in humans norfluoxetine (**LVII**) also acts

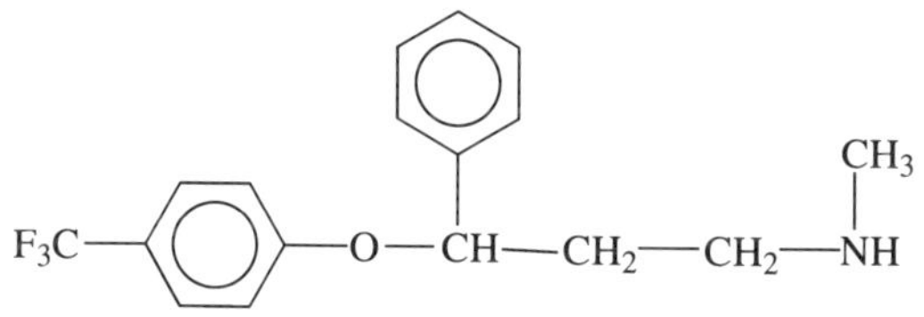

F_3C—⟨benzene⟩—O—CH—CH$_2$—CH$_2$—NH—CH$_3$

(**LVI**)

F_3C—⟨benzene⟩—O—CH—CH$_2$—CH$_2$—NH$_2$

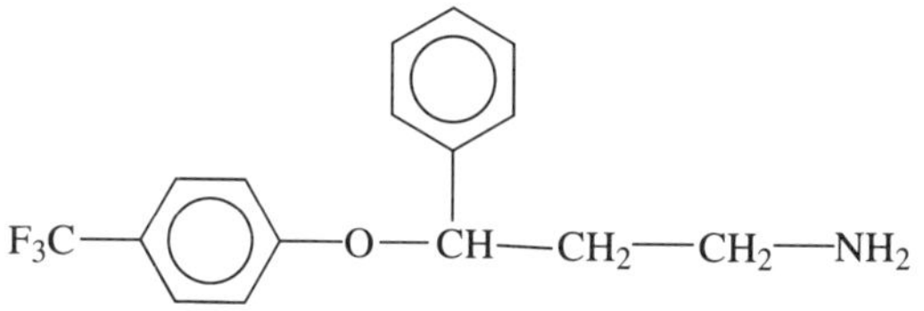

(**LVII**)

similarly. It was approved by the FDA in the USA in 1989 and is favoured there by one in five people on antidepressants. It has been hailed by Kramer[134] as a designer drug to promote assertive behaviour, particularly amongst women, apart from its antidepressant properties. It has been generally prescribed in the UK when the patient has had an adverse reaction to one of the older tricyclic antidepressants. Its use in the USA, the UK and other countries is on the increase. Prozac has a wider margin of safety than the tricyclics, is relatively free of side-effects but is considerably more expensive at ca £1 per capsule as opposed to 1p for the tricyclics. The demand for serum fluoxetine assay is increasing with the increased use of the drug in order to monitor its absorption, distribution and metabolism in a greater number and category of patients in addition to monitoring unexpected toxic concentrations after chronic use of **LVI**.

Fluoxetines have been determined in plasma by gas chromatography with electron-capture detection at therapeutic concentrations,[135] and with flame ionisation or nitrogen-selective detection at toxic concentrations.[136] There is an increasing trend to use HPLC for monitoring therapeutic concentrations of antidepressants[137] with a particular view to improving the specificity of the assays and their LODs. In this case a fluorescence detector provided significantly greater specificity towards fluoxetines than a UV detector. When the latter detector was used at 227 nm, a number of commonly prescribed antidepressants, their metabolites and benzodiazepines eluted close to **LVI** and **LVII**. In addition, the fluorescence detector operated at an excitation wavelength of 235 nm and an emission wavelength of 310 nm gave a lower LOD than the UV detector and could detect 20 ng cm^{-3} of either fluoxetine using 0.5 cm^3 of the sample.

6.7.3 Procedure

A reversed-phase liquid chromatographic procedure with fluorescence detection for the simultaneous determination of fluoxetine and its active metabolite, norfluoxetine, in human serum is described.[133] A 0.5-cm^3 aliquot of the sample after the addition of protriptyline as the internal standard is passed through a 1-cm^3 BondElut C$_{18}$ silica extraction column. The column is selectively washed to remove polar, neutral, acidic and weakly basic compounds. The desired compounds are eluted with a 0.25-cm^3 aliquot of 0.1 mol dm^{-3} perchloric acid–acetonitrile (1 + 3). A 20-μl aliquot of the eluate is injected on to a 15 cm × 4.6 mm i.d. column packed with 5-μm C$_8$-bonded silica particles, which is operated at ambient temperature with a mobile phase of acetonitrile (375 cm^3)–H$_2$O (625 cm^3)–tetramethylammonium perchlorate (1.5 g)–70% perchloric acid (0.1 cm^3). The peaks are detected with a fluorescence detector (excitation at 235 nm; emission at 310 nm). In the resulting chromatogram, there are only a few extraneous peaks and fluoxetines give sharp peaks which are well resolved from peaks of the solvent and internal standard. The extraction recovery of fluoxetines and internal standard is of the order of 85%.

Figure 52A shows the chromatogram of extracted drug-free serum with the well resolved and fluorescent internal standard protriptyline (a rarely prescribed antidepressant) eluted at just over 5 min. A serum standard of 250 ng cm^{-3} of each fluoxetine is given in Figure 52B, again showing baseline resolution of **LVI** and **LVII**. Figure 52C corresponds to a patient receiving fluoxetine which, by the internal standard method of quantitation as discussed in Section 2.8, gives concentrations of 260 ng cm^{-3} for **LVI** and 80 ng cm^{-3} for **LVII**. Figure 52D and E show that the commonly prescribed antidepressants nortriptyline and amitriptyline do not give fluorescence signals but are detected by a UV detector and have an elution time very close to that of **LVII**.

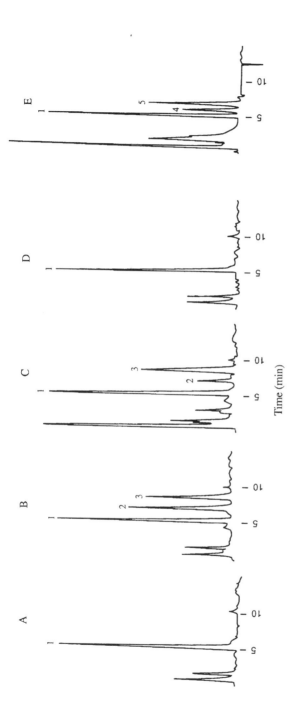

Figure 52 Chromatograms of 20-μl injections of extracts of serum. Fluorescence detection: (A) drug-free serum; (B) serum standard of 250 ng cm^{-3} of each fluoxetine; (C) serum of a patient receiving fluoxetine (norfluoxetine = 80 ng cm^{-3}, fluoxetine = 260 ng cm^{-3}); (D) serum standard of 400 ng cm^{-3} each of nortriptyline and amitriptyline. Absorbance detection at 227 nm: (E) serum standard of 400 ng cm^{-3} each of nortriptyline and amitriptyline. Peaks: 1 = protriptyline; 2 = norfluoxetine; 3 = fluoxetine; 4 = nortriptyline; 5 = amitriptyline. Reproduced by permission of Marcel Dekker Inc from R. N. Gupta and M. Steiner, *J. Liq. Chromatogr.*, 1990, **13**, 3785

6.8 DIRECT DETERMINATION OF THIOAMIDE DRUGS IN BIOLOGICAL FLUIDS BY CATHODIC STRIPPING VOLTAMMETRY[138]

6.8.1 Summary

A method has been developed to determine several sulphur-containing drugs in plasma and urine at concentrations down to 2×10^{-8} mol dm^{-3} using cathodic stripping voltammetry at a hanging mercury drop electrode. Naturally occurring sulphur compounds such as glutathione, thiamine, methionine, cystine, cysteine and hydrogen sulphide and anionic species such as chloride ion are shown not to interfere. Results are shown to be accurate, with a precision of 2.9% at the 5×10^{-7} mol dm^{-3} level on 2-cm^3 samples. An experimental procedure is described which involves a $1+1$ dilution of the sample with Britton–Robinson buffer (pH 4.78) followed by electrolysis at $+0.05$ V (vs SCE) and subsequent cathode stripping to -0.75 V (vs SCE). The standard addition method is used to determine the drug concentration.

6.8.2 Introduction

Thioamide drugs such as **LVIII** are labile to the extent that they can degrade to the corresponding amide or nitrile with the production of elemental sulphur and H_2S during solvent extraction prior to the determination step[139] (e.g. gas–liquid chromatography[140] or high-performance liquid chromatography[141]). A direct technique for their determination in biological fluids at concentrations of the order of 10^{-7}–10^{-8} mol dm^{-3} is therefore of particular importance and such a technique should discriminate between the drug, it metabolites and naturally occurring compounds in the biological fluid. For thioamides such as **LVIII** there exists the possibility of electrolytically plating the drug on to a mercury indicator electrode as a mercury compound under anodic conditions with subsequent quantitation by cathodic stripping voltammetry. Since thioamides such as ethionamide can metabolise to non-sulphur-containing metabolites such as the amide and carboxylic acid,[142] such an analytical technique would be specific for the parent compound in the presence of such metabolites.

[Structure with CSNH$_2$ group] + Hg^{2+} ⟶ [Structure with CN group] + HgS + 2H$^+$

(**LVIII**)

6.8.3 Procedure

To 2.0 cm^3 of plasma or urine sample (containing a maximum of 10^{-6} mol dm^{-3} of drug) in the polarograph cell were added 2.0 cm^3 of Britton–Robinson buffer (pH 4.78). The solution was carefully degassed by bubbling oxygen-free nitrogen through the solution for 5 min, keeping frothing to a minimum. The cell was attached to the electrode assembly and a nitrogen flow maintained over the solution. A five-division mercury drop was dialled on the working electrode and the magnetic stirrer was switched on. The solution was electrolysed for 5 min (accurately timed with a stop-watch) at +0.05 V, the stirrer switched off and electrolysis continued on the quiescent solution for a further 60 s. The potential scan was then performed at 10 mV s^{-1} to −0.75 V to obtain a cathodic stripping voltammogram (Figure 53b). The scan direction was immediately reversed, with stirring, back to +0.05 V and the plating/stripping process repeated. The mean of the two signals was then taken for this particular concentration.

At the end of the last scan, the potential was held at −0.75 V, 0.10 cm^3 of a 5 × 10^{-6} mol dm^{-3} solution of the drug in buffer solution added and a reverse scan back to +0.05 V performed, with stirring. The plating/stripping process was then repeated as above in duplicate. A further 0.10-cm^3 standard addition was then similarly made and the cyclic process again repeated in duplicate. Sample concentrations were calculated in the usual manner for standard additions.

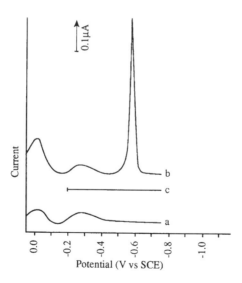

Figure 53 Cathodic stripping voltammograms for 1+1 horse plasma–pH 4.78 buffer solutions containing (a) no drug, (b) 3.49 × 10^{-7} mol dm^{-3} compound **LVIII**. Scan (c) represents a post-cathodic stripping scan of solution (b). Reprinted with permission from I. E. Davidson and W. F. Smyth, *Anal. Chem.*, 1977, **49**, 1195. Copyright 1977 American Chemical Society

Blank scans were run on samples taken prior to drug administration to ensure the absence of polarographic activity in the vicinity of the drug stripping peak (Figure 53a). Figure 53c shows a rescan of (b) from -0.2 V immediately after a stripping procedure, showing all plated material to have been removed in a single scan.

Possible interferences include Cl^- and naturally occurring sulphur compounds such as glutathione, methionine, thiamine, cystine and cysteine. Figure 54a, b and c show the effect of successive additions of, respectively, cysteine, reduced glutathione and chloride ion at concentrations that would be encountered in plasma, on the stripping peak of 3.7×10^{-7} mol dm^{-3} **LVIII** in horse plasma–pH 4.78 buffer $(1+1)$. The peak height is seen to be unaffected by these possible interferents. The broad peaks of cysteine and reduced glutathione can be seen from -0.3 to -0.5 V.

The reason for this electrochemical discrimination of **LVIII** from its potential interferences is the different plating mechanism for **LVIII** and related compounds, i.e. the plated compound is HgS and not RSHg for the potentially interfering thiols as illustrated in the Introduction.

This technique is found to give linear calibration in the range 5×10^{-8}–10^{-6} mol dm^{-3} with an LOD of 2×10^{-8} mol dm^{-3}. The precision is 2.9% at the

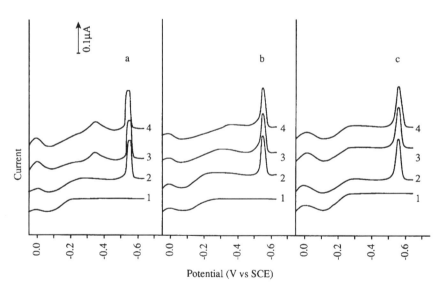

Figure 54 Effect of possible interferents on stripping voltammograms for $1+1$ horse plasma–pH 4.78 buffer solutions with 3.7×10^{-7} mol dm^{-3} compound **LVIII**. (a) Additions of (2) 0.0, (3) 12.0 and (4) 24.0 µg of cysteine; (b) additions of (2) 0.0, (3) 3.0 and (4) 6.0 µg of reduced glutathione; (c) additions of (2) 0.0, (3) 15.0 and (4) 30.0 mg of sodium chloride. Scans (1) in each case are for blanks with no drugs or additives. Reprinted with permission from I. E. Davidson and W. F. Smyth, *Anal. Chem.*, 1977, **49**, 1195. Copyright 1977 American Chemical Society

5×10^{-7} mol dm^{-3} level on 2-cm^3 samples. In addition, non-sulphur-containing metabolites and degradation products of **LVIII** do not interfere.

6.8.4 Calculation

The thioamide drug **LVIII** gave a cathodic stripping voltammogram at -0.55 V (vs SCE) using a hanging mercury drop electrode and a supporting electrolyte of blank plasma–pH 4.78 Britton–Robinson buffer (1 + 1). A 2-cm^3 volume of a plasma sample containing an unknown concentration of **LVIII** was mixed with 2-cm^3 of pH 4.78 Britton–Robinson buffer in the voltammetric cell, deaerated and subjected to cathodic stripping voltammetry with deposition at $+0.05$ V for 2 min to give a peak at -0.55 V (vs SCE) of height 0.50 μA. A 0.10 cm^3 volume of a 5×10^{-6} mol dm^{-3} standard solution of **LVIII** was then added to the voltammetric cell and the above process repeated to give a peak at an identical potential of height 0.80 μA. Calculate the concentration of **LVIII** in the original plasma sample.

Using the standard addition equation given in Section 4.1.4:

$$C_u = \frac{0.50 \times 0.10 \times 5 \times 10^{-6}}{(0.30 \times 4) + (0.80 \times 0.10)}$$

$$= \frac{0.50 \times 0.10 \times 5 \times 10^{-6}}{1.28}$$

$$= 1.9 \times 10^{-7} \text{ mol dm}^{-3}$$

Therefore the concentration in the original plasma sample was 3.8×10^{-7} mol dm^{-3}.

Chapter 7

Analysis of High Molecular Weight Analytes

7.1 ANALYSIS OF ABNORMAL OR VARIANT HAEMOGLOBINS BY HPLC–ELECTROSPRAY MS

7.1.1 Summary

HPLC–electrospray mass spectrometry can be used for the analysis of abnormal or variant haemoglobins. Samples of 10 pmol are injected into an HPLC mobile phase of equal parts of water and methanol containing 0.1% formic acid followed by electrospray mass spectrometric detection. Haemoglobins give a series of multiply charged ions which can be transformed by appropriate software to give a single peak for each haemoglobin (α, β and γ). Abnormal or variant haemoglobins give rise to different mass spectral displays for α, β and γ haemoglobins when compared with normal human haemoglobins and this can be used to diagnose illnesses such as sickle cell anaemia.

7.1.2 Introduction

Electrospray mass spectrometry may be applied to the determination of the relative molecular mass (RMM) of large molecules because of the tendency for multiply charged ions to be formed. Mass spectrometers display spectra according to the mass to charge ratio (m/z) and, for example, a molecule of RMM 30 000 having 20 charges gives a peak near m/z 1500. In electrospray, ions may be highly protonated: the ion with 20 positive charges would have a composition of $(M+20H)^{20+}$ and hence the corresponding peak would appear at $(M+20)/20$, i.e. m/z 1501. Hence mass spectra from molecules whose RMM would greatly exceed the range of normal sector and quadrupole instruments can be observed. A typical positive-ion electrospray mass spectrum of myoglobin (RMM 16 951.5) demonstrates this point (Figure 55). An envelope of peaks can be seen corresponding to the intact molecule carrying different numbers of charges ranging from 23 at m/z 738 to 12 at m/z 1413. When the RMM is unknown it can be calculated by application of two simple

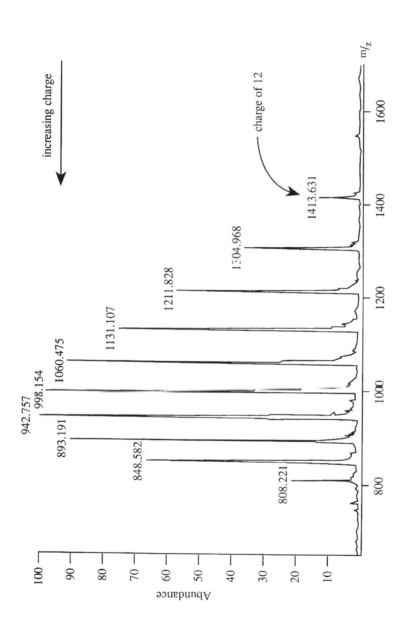

Figure 55 Positive-ion electrospray mass spectrum of myoglobin, RMM 16951.5. Reproduced by permission of VG Instruments from *VG Monographs in Mass Spectrometry*, No. 3

equations. For two adjacent peaks, m/z (m_1) and m/z (m_2), where $m_1 < m_2$:

$$n_2 = \frac{m_1 - H}{m_2 - m_1} \tag{43}$$

$$M = n_2(m_2 - H) \tag{44}$$

where $n_2 =$ the number of charges on the ion at m/z (m_2), $H =$ RMM of one proton, and $M =$ RMM of the unknown.

7.1.3 Procedure

An interesting application using HPLC–electrospray MS has been described for the analysis of haemoglobins.[143] Figure 56 shows part of the positive-ion electrospray mass spectrum of normal human haemoglobin (HbA). The molecule consists of four polypeptide chains around a central core of haem. There are two pairs of chains called α- and β-globin which can be clearly observed as a series of multiply charged ions (A, B and C, Figure 56a). Here the electrospray data are presented as a series of multiply charged ions with a textual output of the actual RMM. However, transformation software exists that presents the data as a reconstructed mass spectrum as if

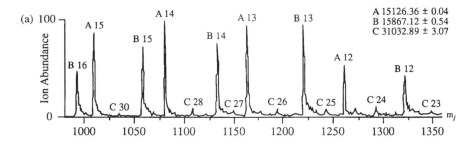

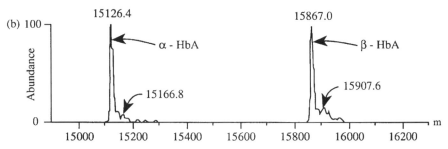

Figure 56 Positive-ion electrospray mass spectrum of normal human haemoglobin: (a) the envelope of multiply charged ions originally produced; (b) the transformed data showing clearly the α- and β-chain masses. Reproduced by permission of VG Instruments from *VG Monographs in Mass Spectrometry*, No. 3

the sample had yielded a singly charged molecular ion (Figure 56b). The RMMs of the two principal components can be simply read off as α-HbA (15 126.4; theoretical value 15 126.4) and β-HbA (15 867.0, theoretical value 15 867.2). The mass measurement accuracy of electrospray is 0.01% or better of the theoretical value and the results from HbA easily fall within these limits.

Electrospray MS has been applied to the analysis of abnormal or variant haemoglobins.[143-145] Figures 57 and 58 illustrate the data that can be achieved. All analyses were performed by injecting 10-pmol samples into an HPLC mobile phase of equal parts of water and methanol containing 0.1% of formic acid. Figure 57a

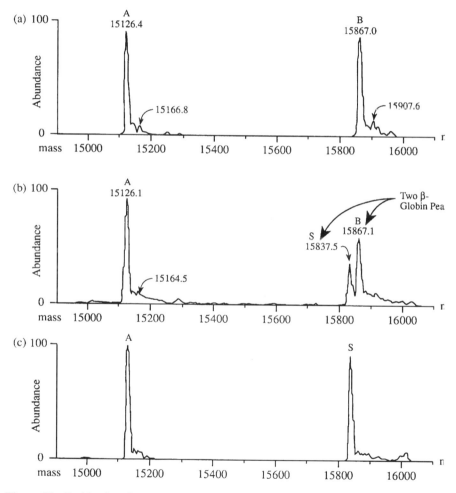

Figure 57 Positive-ion electrospray mass spectra of adult human haemoglobins: (a) normal adult; (b) adult heterozygous for the sickle cell anaemia gene; (c) adult homozygous for the sickle cell anaemia gene. Reproduced by permission of VG Instruments from *VG Monographs in Mass Spectrometry*, No. 3

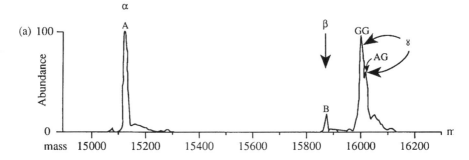

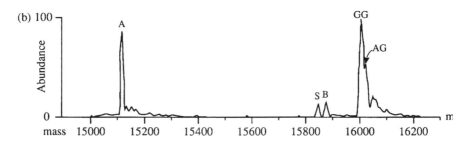

Figure 58 Positive-ion electrospray mass spectra of foetal haemoglobin samples: (a) normal foetus; (b) foetal sample from a subject heterozygous for the sickle cell anaemia-causing gene. Reproduced by permission of VG Instruments from *VG Monographs in Mass Spectrometry*, No. 3

shows peaks for α- and β-globin from normal HbA but Figure 57b obtained from an adult human shows two β-globin peaks, labelled S and B and mass measured at 15 837.5 and 15 867.1, respectively. This indicates that the human subject is heterozygous for the sickle cell anaemia gene. The presence of the second β-globin (S) arises from the substitution of glutamic acid by valine at position 6 of the β-chain. Thus the sample contains both normal and abnormal β-chains and the subject is a carrier for sickle cell anaemia. Figure 57c shows a homozygous case with all of the β-globin present in the sickle form. This subject is affected by sickle cell anaemia and would be expected to show clinical symptoms. The single β-chain is easily observed and its shift in mass from the normal HbA β-chain is clearly revealed.

In Figure 58a and b, a slightly more complex situation is illustrated. Foetal haemoglobin (HbF) has a different structure to that of adults, in containing α-, β- and γ-globin chains, with the γ form being composed of major components GG and AG. The transformed electrospray mass spectrum of HbF (Figure 58a) shows peaks corresponding to the three globins. In Figure 58b, an extra β-globin is clearly observable, again indicating a subject who is heterozygous for sickle cell anaemia and demonstrating that HPLC–electrospray MS can again be used in diagnosis.

7.2 DETERMINATION OF INSULIN BY RADIOIMMUNOASSAY (RIA)

7.2.1 Summary

The β-emitter ^{125}I can be inserted into the amino acid tyrosine in insulin and this labelled antigen allowed to compete with unlabelled insulin for a limited amount of antibody, thus giving rise to a radioimmunoassay for the determination of insulin directly in small volumes (e.g. 50 μl) of biological fluids and at concentrations normally encountered in these situations, i.e. low ng cm^{-3} levels.

7.2.2 Introduction

Radioimmunoassay (RIA) was initially developed in 1962 for the determination of a naturally occurring immunogen or antigen, the hormone insulin, and is now very widely applied to the determination of a wide range of organic molecules such as steroids, vitamins and drugs in biological fluids. The essential equations used in the operation of RIA are as follows, illustrating the competition of Ag* and Ag for a limited amount of Ab:

$$Ag^* + Ab \rightleftharpoons Ag^* - Ab \qquad (45)$$
$$Ag + Ab \rightleftharpoons Ag - Ab \qquad (46)$$

Ag* and Ag represent the radiolabelled and unlabelled (to be determined) antigen, respectively, and Ab represents the antibody or antiserum, which is usually a γ-globulin or immunoglobulin which reacts specifically with the antigen. Ab is prepared by injecting Ag into the blood of a test animal, resulting in production of Ab in the animal's blood, which is then sampled and Ab prepared accordingly. Ag* is prepared by using a radioisotope such as ^{125}I ($t_{\frac{1}{2}} = 60$ days, $E_\gamma = 0.035$ MeV) or ^{3}H ($t_{1/2} = 12.26$ years, $E_\beta = 0.015$ MeV) which can be inserted in Ag without too much difficulty. ^{125}I is particularly used because it can be inserted into appropriate organic entities such as the tyrosine amino acid (**LIX**) in insulin and other molecules.

HO—⟨◯⟩—CH$_2$—CH(—$^+$NH$_3$)—COO$^-$ →[Chloramine-T / NaI*/pH 8]→ HO—⟨◯⟩(I*)(I*)—CH$_2$—CH(—$^+$NH$_3$)—COO$^-$

(**LIX**)

In addition, ^{125}I is a γ-emitter and is thus readily counted on a scintillation counter. Its half-life of 60 days means that it can be used for almost a year before decay reduces its radioactivity to insignificant levels.

Following equilibration as in equations 45 and 46, Ag*–Ab and Ag–Ab complexes are separated from free Ag* and Ag using techniques such as precipitation by $(NH_4)_2SO_4$ followed by centrifugation, electrophoresis or adsorption on dextran-coated charcoal. The radioactivity remaining in solution relative to that bound in the complex is then measured and related to a calibration curve as illustrated in Sections 7.2.3 and 7.2.4.

Most radioimmunoassays are carried out using commercially available screening kits, use small sample volumes (e.g. 50 µl) and are well suited to the task of screening large numbers of samples. A wide range of molecules can be determined using ^{125}I screening kits, e.g. hormones such as insulin and many others, vitamins and drugs such as digitoxin, digoxin, gentamicin, morphine, phenobarbital, phenytoin and theophylline. Molecules determined with ^{3}H or ^{14}C kits include morphine, barbiturates, marijuana, LSD, vitamin D_3 and many hormones. Radioimmunoassays can display cross reactivities to other related molecules, e.g. a wide range of barbiturates and their hydroxylated metabolites will all bind to a particular antiserum, although potentially interfering molecules such as primidone, phenytoin, caffeine and theophylline show a very low level of binding.[31] This has been further discussed in Section 2.2. This section, however, deals with the determination of low ng cm^{-3} concentrations of insulin in biological fluids such as human blood. Insulin has a molecular weight of ca 6000 and is a two-chain polypeptide.

7.2.3 Procedure

(a) To a known volume of the biological fluid containing the unknown concentration of insulin is added a known volume of solution of labelled insulin of known radioactivity.
(b) A known but limited quantity of the appropriate Ab is then added.
(c) Equilibration is achieved; 10 min would not be an uncommon equilibration time.
(d) Ag*–Ab and Ag–Ab are separated and the radioactivity remaining in solution is measured relative to that of the separated complexes to give a value of

$$\frac{\text{radioactivity of Ag}^*}{\text{radioactivity of Ag}^*\text{-Ab}}$$

(e) This value is then referred to a preconstructed calibration curve to give the unknown concentration of insulin in the biological fluid.

7.2.4 Calculation

An RIA calibration curve for insulin determination was prepared by mixing 50 µl of standard insulin solutions in the concentration range 3–9 ng cm^{-3} each with 100 µl of labelled insulin solution. The total radioactivity of each solution was 2×10^4 counts min^{-1}. The same but limited quantity of antibody was added to each solution,

Analysis of High Molecular Weight Analytes

Table 37 Radioimmunoassay data for standards and sample in the determination of insulin in 50 µl of a biological fluid

	(ng cm^{-3})				
[Insulin] Concentration	3.0	5.0	7.0	9.0	Unknown
Radioactivity of complexes (counts min^{-1})	13245	11111	9852	9091	10100
Radioactivity of free Ag* (counts min^{-1})	6755	8889	10148	10909	9900
$\dfrac{\text{Radioactivity of Ag}^*}{\text{Radioactivity of Ag}^*-\text{Ab}}$	0.57	0.80	1.03	1.20	0.98

the mixtures were equilibrated and in each case the complexes were separated and the radioactivity measured. The same procedure was then followed using 50 µl of a biological fluid containing an unknown insulin concentration. The results are presented in Table 37.

A plot of radioactivity of Ag*/radioactivity of Ag*–Ab vs [insulin] in ng cm^{-3} will be curved and from it, it can be found that the biological fluid will have an insulin concentration of 6.5 ng cm^{-3}. Alternatively, the calibration plot can be prepared as percentage of bound activity or percentage of free activity vs [insulin] in ng cm^{-3}.

7.3 FLUORESCENCE OF EUROPIUM CHELATES AND ITS APPLICATION TO FLUOROIMMUNOASSAY OF PROSTATE-SPECIFIC ANTIGEN (PSA) IN HUMAN SERUM

7.3.1 Summary

Lanthanides such as europium, in the form of certain chelates, give characteristic fluorescence at relatively high wavelengths (> 500 nm). This can be put to use in automated solid-phase, two-site fluoroimmunoassay based on the direct sandwich technique for the determination of prostate-specific antigen (PSA) in human serum. PSA is first reacted with immobilised monoclonal antibodies directed against a specific antigenic site on the PSA. Eu^{3+}-labelled antibodies directed against a separate antigenic site are then reacted with the PSA already bound to the solid-phase antibody. Finally, Eu^{3+} is dissociated from the labelled antibody and its fluorescence, in chelate form, is proportional to the concentration of PSA in the sample. Linearity is achieved in the range 1–500 µg dm^{-3} of PSA with an LOD of the order of 0.1 µg dm^{-3}. The results provide a strong indication of prostatic cancer.

7.3.2 Introduction

There are a limited number of simple inorganic species that are naturally fluorescent in aqueous solution and they are certain ions of the lanthanides (rare earths), e.g. Sm^{3+}, Eu^{3+}, Gd^{3+}, Tb^{3+} and Dy^{3+}, and actinides (transuranics) and several others such as Tl$^+$ in the presence of excess Cl$^-$. The fluorescence emission of these lanthanides is weak but is significantly enhanced in the form of certain chelates, e.g. europium salicylate. Such chelates give characteristic 'line' emissions at relatively high wavelengths (> 500 nm) with Stoke shifts of 250–300 nm (Figure 59). The line emissions are due to transitions involving non-bonding electrons in the atomic f-orbitals, not a triplet state of the overall chelate molecule. This high wavelength emission is of particular value when these fluorescent entities are complexed with biological compounds such as proteins in immunoassays, since other constituents of the biological sample will fluoresce at significantly lower wavelengths in the ultraviolet region of the spectrum. Furthermore, lanthanide chelates have long fluorescence lifetimes of the order of 0.5–1.0 ms, and this has been put to use in time-resolved fluorimetry to discriminate high background scatter (decay time ca 10 ns) and background fluorescence. In this case, the label (the lanthanide chelate) is pulsed

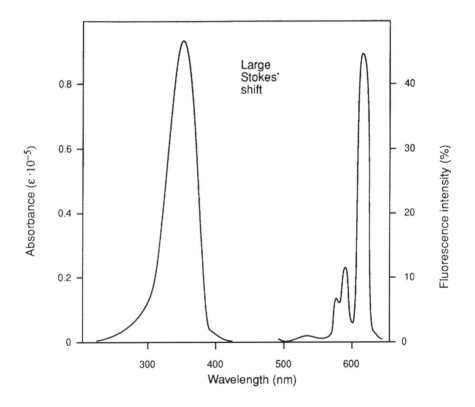

Figure 59 Fluorescence of lanthanide chelates characterised by a large Stokes' shift and a narrow emission peak. Reprinted from *American Laboratory*, volume 24, number 1, page 14, 1994. Copyright 1994 International Scientific Communications, Inc.

1000 times per second with excitation light of 340 nm. In the period between the flashes, the specific fluorescence is measured after the appropriate delay time (e.g. after 400 µs in the case of Eu^{3+} chelates). Statistically accurate results are therefore achieved using a short measuring time of 1 s. Eu^{3+} is the most commonly used of the tripositive lanthanides and in particular as a label in chelate form for antibodies in solid-phase immunoassays,[146] e.g. N^1-(p-isothiocyanatobenzyl)diethylenetriamine-N^1,N^2,N^3,N^3-tetraacetic acid chelated with Eu^{3+} (**LX**). It is the aromatic isothiocyanato group that readily reacts with a free amino group on the protein to form a thiourea derivative (**LXI**). Antibodies with as many as 10–15 Eu^{3+} per immunoglobulin G (IgG) molecule can be prepared in this way.

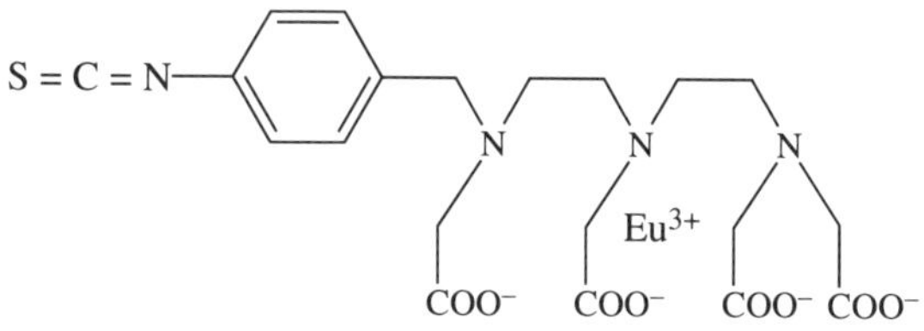

(**LX**)

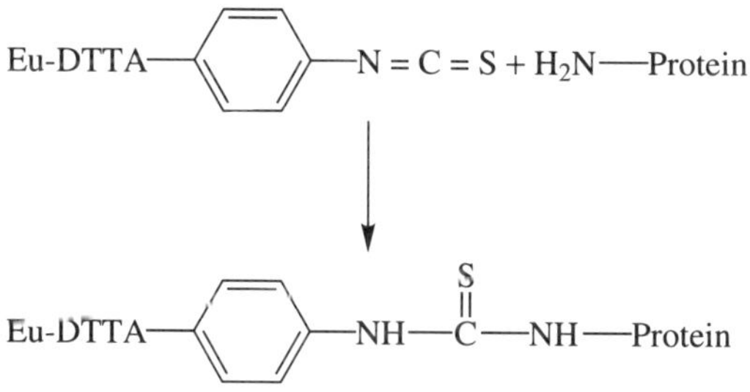

(**LXI**)

Diagnostic kits are now available for a wide range of analytes for major diagnostic areas such as thyroid function, fertility, oncology and pre- and neonatal screening for operation in automated instrumentation such as the DELFIA system (Wallac Oy, Turku, Finland). As an example, this DELFIA system can be used to quantitate prostate-specific antigen (PSA) in human serum using the Auto DELFIA PSA kit. PSA is a glycoprotein with a molecular weight of 34 000 and a carbohydrate content of 7% and is believed to be serine protease. Events in the prostate such as malignancy and others can lead to elevated serum PSA levels, so its determination in this complex matrix is of value in screening for prostatic cancer.

7.3.3 Procedure

The Auto DELFIA PSA assay[147] is an automated solid-phase, two-site fluoroimmunometric assay based on the direct sandwich technique in which two monoclonal

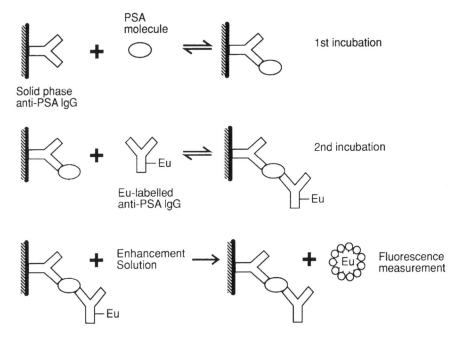

Figure 60 Reaction scheme in fluoroimmunoassay of prostate-specific antigen. Reproduced by permission of Wallac Oy from *Application Note on the Fluoroimmunoassay of Prostate-Specific Antigen (PSA) in Human Serum*

antibodies (derived from mice) are used. Standard, control and patient samples (blood is collected by venipuncture, allowed to clot and serum is separated by centrifugation) containing PSA undergo a series of reactions as illustrated in Figure 60.

First PSA is reacted with immobilised monoclonal antibodies directed against a specific antigenic site on the PSA. Europium-labelled antibodies directed against a separate antigenic site are then reacted with the PSA already bound to the solid-phase antibody. Finally an enhancement solution of low pH is added, which dissociates Eu^{3+} ions from the labelled antibody into solution, where they form highly fluorescent chelates with components of the enhancement solution inside protective micelles.

The fluorescence from each sample is proportional to the concentration of PSA in the sample. Fluorescence is linearly related to concentration of PSA in the range 1–500 $\mu g\ dm^{-3}$ with a correlation coefficient of 0.995. A detection limit of the order of 0.1 $\mu g\ dm^{-3}$ has been reported. Elevated PSA levels of $> 7\ \mu g\ dm^{-3}$ are rarely found in non-prostatic malignancies, so this automated fluoroimmunoassay provides a strong indication of prostatic cancer but cannot be used as an absolute screening method for this disease state.

References

1. Locke D. C. and Grossman, W. E. L. (1987) *Anal. Chem.* **59**, 829A.
2. Bowen, H. J. M. (1979) *Environmental Chemistry of the Elements*, Academic Press, London.
3. Szabadvary, F. (1966), *The History of Analytical Chemistry*, Pergamon Press, Oxford.
4. Clark, K. *Civilisation*, p. 7.
5. Szabadvary, F. (1975) *Period, Polytech.*, **19**, 339.
6. Maddison, R. E. W. (1963) *Notes Rec. R. Soc.*, **18**, 104.
7. Maddison, R. E. W. (1958) *Notes Rec. R. Soc.*, **13**, 128.
8. Boyle, R. (1772) *The Works ... in Six Volumes. To Which is Prefixed a Life of the Author*, new edition, J. & F. Rivington et al., London, Vol. 6, p. 54.
9. Stephen, W. I. (1979) *Proc. Anal. Div. Chem. Soc.*, **16**, 91.
10. Belcher, R. (1976) *Anal. Chim. Acta*, **86**, 1.
11. Belcher, R. (1977) *Proc. Anal. Div. Chem. Soc.*, **14**, 161.
12. Belcher, R. (1979) *Proc. Anal. Div. Chem. Soc.*, **16**, 175.
13. West, R. C. (Ed.) (1975–76) *CRC Handbook of Chemistry and Physics*, 56th edn, CRC Press, Cleveland, OH, p. C617.
14. Sandell, E. B. and West, T. S. (1979) *Pure Appl. Chem.*, **51**, 43.
15. *Peak, Hewlett-Packard J.* (1994), **1**, 1.
16. Miranda Ordieres, A. J., Costa Garcia, A., Tunon Blanco, P. and Smyth, W. F. (1987) *Anal. Chim. Acta*, **202**, 141.
17. BP Veterinary, 1977, HMSO publication.
18. Smyth, W. F. (Ed.) (1979) *Polarography of Molecules of Biological Significance*, Academic Press, London, p. 276.
19. Lewis, A. C., Seakins, P. W., Denha, A. M., Bartle, K. D. and Pilling, M. J. (1995) *Atmospheric Environment*, **29**(15), 1871.
20. Thompson, M. and Wood, R. (1993) *Pure Appl. Chem.*, **65**, 2123.
21. Patey, A. L. (1994) *VAM Bull.*, **11**, 12.
22. Howitz, W. and Albert, R. (1987) *Anal. Proc.*, **24**, 49.
23. Morrison, G. H. (1979) *CRC Crit. Rev. Anal. Chem.*, **8**, 296.
24. Pickett, E. E. and Koirtyohann, S. R. (1969) *Anal. Chem.*, **41**(14), 28A.
25. Miller, J. C. and Miller, J. N. (1988) *Statistics for Analytical Chemistry*, 2nd edn, Ellis Horwood, Chichester.
26. ACS Committee on Environmental Improvement and Subcommittee on Environmental Analytical Chemistry (1980) *Anal. Chem.*, **52**, 2242.
27. Analytical Methods Committee (1987) *Analyst*, **112**, 199
28. Kohn, A. (1994), *LCGC Int.* **7**, 652.
29. Oxspring, D. A., O'Kane, E., Marchant, R. and Smyth W. F. (1994), *Anal. Methods Instrum.*, **1** (4), 19.

30. Dennis, B. L., Moyers, J. L. and Wilson, G. S. (1976) *Anal. Chem.*, **48**, 1611.
31. Mason, P. A., Law, B., Pocock, K. and Moffat, A. C. (1982) *Analyst*, **107**, 629.
32. Kratochvil, B. and Taylor, J. K. (1981) *Anal. Chem.*, **53**, 924A.
33. Whitaker, T. B. (1977) *Pure Appl. Chem.*, **49**, 1709.
34. Manahan, S. E. (1986) *Quantitative Chemical Analysis*, Brooks/Cole, p. 634.
35. Shanahan, I. (1992) in *Chemical Analysis in Complex Matrices* ed. M. R. Smyth, Ellis Horwood, Chichester, p. 192.
36. *SKC Comprehensive Catalogue and Air Sampling Guide* (1994–95), SKC.
37. Mitchell, R. L. and Scott, R. O. (1948) *Spectrochimica Acta*, **3**, 367.
38. Cooper, M. J., Sinarko, A. R., Anders, M. W. and Mirkin, B. L. (1976) *Anal. Chem.*, **48**, 1110.
39. Parker, C. R. (1972) *Water Analysis by Atomic Absorption*, Varian Techtron, Australia.
40. Simpson, N. (1992) *Int. Chromatogr. Lab.*, **11**, 7.
41. Skoog, D. A., West, D. M. and Holler, F. J. (1992) *Fundamentals of Analytical Chemistry* 6th edn, Saunders College Publishing.
42. Fifield, F. W. and Kealey, D. (1990) *Principles and Practice of Analytical Chemistry*, 3rd edn, Blackie, Glasgow.
43. Manahan, S. E. (1986) *Quantitative Chemical Analysis*, Brooks/Cole, CA.
44. Kealey, D. (1986) *Experiments in Modern Analytical Chemistry*, Blackie, Glasgow.
45. Smyth, W. F. and Chabala, E. D. (1993). *Fresenius. J. Anal. Chem.*, **345**, 701.
46. McGrath, G., McClean, S., O'Kane, E., Smyth, W. F. and Tagliaro, F. (1995) *J. Chromatogr.* in press.
47. Silverstein, R. M., Bassler, G. C. and Morrill, T. C. (1991) *Spectrometric Identification of Organic Compounds*, Wiley, New York, p. 3.
48. Oxspring, D. A., Smyth, W. F. and Marchant, R. (1995) *Analyst*, **120**, 1995.
49. *Metrohm Application Bulletin*, No. 147e Metrohm, Herisau.
50. Weissler, A. (1945) *Ind. Eng. Chem., Anal. Ed.*, **17**, 695.
51. Van den Berg, C. M. G. and Huang, A. A. (1984) *Anal. Chem.*, **56**, 2383.
52. Kopp, J. F. and Kroner, R. C. (1967). *Trace Metals in Waters of USA*, US Department of the Interior: FWPCA, Division of Pollution Surveillance, Cincinnati, OH.
53. Harvey, M. W. (1963) *Chemistry and Fertility of Sea Water*, Cambridge University Press, Cambridge.
54. Chau, Y. K. and Lum-Shue-Chau, K. (1969) *Anal. Chim. Acta*, **48**, 205.
55. Smith, W. and Smith, A. (Eds) (1975) *Minamata*, Holt Rinehart and Winston, New York.
56. Jensen, S. and Jernelov, A. (1969) *Nature*, **223**, 753.
57. Rodriguez-Vasquez, J. A. (1978) *Talanta*, **25**, 299.
58. Holak, W. (1982) *Analyst*, **107**, 1457.
59. Edward, R. A. (1980), *Analyst*, **105**, 139.
60. Benjamin, W. (1971) *Social Sci. Plant Anal.*, **2**, 363.
61. Oxspring, D. A., McClean, S., O'Kane, E. and Smyth, W. F. *Anal. Chim. Acta*, in press.
62. Noden, F. G. (1980) in *Lead in the Marine Environment*, ed. M. Brancia and Z. Konrad, Pergamon Press, Oxford, pp. 83–91.
63. Radojevic, M. (1989) in *Environmental Analysis Using Chromatography Interfaced with Atomic Spectroscopy*, ed. R. M. Harrison and S. Rapsomanikis, Ellis Horwood, Chichester, Chapter 8, pp. 223–257.
64. Radojevic, M., Allen, A., Rapsomanikis, S. and Harrison, R. M., unpublished data.
65. Crecelius, E. A. and Sanders, R. W. (1980) *Anal. Chem.*, **52**, 1310.
66. Robbins, W. B. and Caruso, J. A. (1979) *Anal. Chem.*, **51**, 889A.
67. Kadeg, R. D. and Christian, G. D. (1977) *Anal. Chim. Acta*, **88**, 117.
68. Gifford, P. R. and Bruckenstein, S. (1980) *Anal. Chem.*, **52**, 1028.
69. Henry, F. T. and Thorpe, T. M. (1980) *Anal. Chem.*, **52**, 80.

70. Bodewig, F. G., Valenta, P. and Nurnberg, H. W. (1982) *Fresenius' Z. Anal. Chem.*, **311**, 187.
71. Holak, W. (1980) *Anal. Chem.*, **52**, 2189.
72. Ricci, G. R. (1981) *Anal. Chem.*, **53**, 610.
73. De Bourienne, L. A. F. (1885) *Memoires of Napoleon Bonaparte*, Vol. 4, Grolier Society, London, pp. 427–428.
74. Forshufvud, S. (1962) *Who Killed Napoleon?*, Hutchinson, London.
75. Forshufvud, S. and Weider, B. (1978) *Assassination at St Helena*, Mitchell, Vancouver.
76. Weider, B. and Hapgood, D. (1982) *The Murder of Napoleon*, Robson Books, London.
77. Smith, H., Forshufvud, S. and Wassen, A. (1962) *Nature*, **194**, 725.
78. Forshufvud, S., Smith, H. and Wassen, A. (1964) *Arch. Toxicol.*, **20**, 210.
79. Leslie, A. C. D. and Smith, H. (1978) *Arch. Toxicol.*, **41**, 163.
80. Forshufvud, S., Smith, H. and Wassen, A. (1961) *Nature*, **192**, 103.
81. Lewin, P. K., Hancock, R. G. V. and Voynovich, P. (1982) *Nature*, **299**, 627
82. *Analysis Europa*, February 1995, 16.
83. Takeuchi, T. (1982) *Radioanal. Chem.*, **70**, 29.
84. Smith, H. (1959) *Anal. Chem.*, **31**, 1361.
85. Challenger, F. (1945) *Chem. Rev.*, **36**, 315.
86. Hunter, D. (1978) *Diseases of occupation*, Hodder and Stoughton, London, pp. 348, 363.
87. Sanger, C. R. (1893) *Proc. Am. Acad. Arts Sci.*, **29**, 148.
88. Gosio, E. (1893) *Arch. Ital. Biol.*, **18**, 235.
89. Jones, D. E. H. and Ledingham, K. W. D. (1982) *Nature*, **299**, 626.
90. Longwood Old House (Official pamphlet) (1960) 8–9.
91. Brunton, T. L. (1883) *Br. Med. J.*, **1**, 1218.
92. *Analysis Europa*, February 1995, 12.
93. *Analysis Europa*, February 1995, 11.
94. Wells, N. (1967) *N.Z. Geol. Geophys.*, **10**, 198.
95. Fleming, G. A. (1962) *Soil Sci.*, **94**, 28.
96. Beath, O. A. (1962) *Sci. News Lett.*, **81**, 254.
97. Chau, Y. K., Wong, P. T. S., Silverberg, B. A., Luxon, P. L. and Bengert, G. A. (1976) *Science*, **192**, 1130.
98. Reamer, D. C. and Zoller, W. M. (1980) **208**, 500.
99. Cuffer, G. A. and Bruland, K. W. (1984) *Limnol. Oceanogr.*, **29**, 1179.
100. Takayanagi, T. and Wong, W. (1983) *Anal. Chim. Acta*, **148**, 263.
101. Verlinden, M, Deelstra, H. and Adriaenssens, E. (1981) *Talanta*, **28**, 637.
102. Flinn, C. G. and Aue, W. A. (1978) *J. Chromatogr.*, **153**, 49.
103. Olsen, K. B., Sklaren, D. S. and Evans, J. C. (1985) *Spectrochim. Acta*, **40**, 357.
104. Chiang, L., James, B. D. and Magee, R. J. (1989) *Mikrochim. Acta (Wien) II*, 149.
105. Young, J. W. and Christian, G. D. (1973) *Anal. Chim. Acta*, **65**, 127.
106. Ulivo, A. D. and Papoff, P. (1986) *J. Anal. At. Spectrom.*, **1**, 479.
107. Irgolic, K. J., Stockton, R. A., Chakraborti, D. and Beyer, W. (1983) *Spectrochim. Acta, Part B*, **38**, 437.
108. Cutter, G. A. (1978) *Anal. Chim. Acta*, **98**, 59.
109. Cutter, G. A. (1985) *Anal. Chem.*, **57**, 2951.
110. Rendell, D. (1987) *Fluorescence and Phosphorescence*, ACOL Series, Wiley, Chichester, p. 278.
111. Delves, H. T. (1988) *Chem. Br.*, October, 1009.
112. Shearer, E. (1972) *J. Pharm. Sci.*, **61**, 1627.
113. Hollis, D. P. (1963) *Anal. Chem.*, **35**, 1682.
114. Das Gupta, V. (1980) *J. Pharm. Sci.*, **69**, 110.
115. *US Pharmacopea*, 20th Revision, 1980, p. 570.

116. Reif, V. D. and Deangelis, N. J. (1983) *J. Pharm. Sci.*, **72**, 1330.
117. Capillary Electrophoresis Course, Chemistry Department, University of York, September 1993.
118. Abdalla, A. and Fogg, A. (1983) *Analyst*, **108**, 53.
119. Bundgaard, H. (1972) *J. Pharm. Pharmacol*, **24**, 790.
120. Sengun, S. (1986) *Talanta*, **33**, 363.
121. Purnell, C. J., Bagon, D. A. and Warwick, C. J. (1982) in *Analytical Techniques in Environmental Chemistry*, ed. J. Albaiges, Pergamon Press, Oxford, p. 209.
122. Wu, W. S., Stoyanoff, R. E. Szklar, R. S. and Gaind, V. S. (1990) *Analyst*, **115**, 801.
123. Leavitt, R. A., Kells, J. J., Bunkelmann, J. R. and Hollingworth, R. M. *Bull. Environ. Contam. Toxicol.*, **46**, 22 (1991).
124. Millipore Technical Brief (1991) *EnviroGard^{TM} Triazine Plate Kit*, Lit. No. TB088, Millipore, Bedford, MA.
125. Millipore Application (1991) *EnviroGard Test Kits for Pesticide Detection*, Cat. No. EU341/U, Millipore, Bedford, MA.
126. Thurman, E. M. (1990) *Anal. Chem.*, 2043.
127. Bushway, R. J. et al. (1988) *Bull. Environ. Contam. Toxicol.*, **40**, 647.
128. *Official Methods of Analysis of the Association of Official Analytical Chemists*, AOAC, Philadelphia, 1984, p. 184.
129. Singer, S. (1966) *Analyst*, **91**, 790.
130. *Annual Report on Research and Technical Work*, Veterinary Research Laboratories, Belfast, 1982, p. 235.
131. Analytichem Application Note M797 (1990) *Extraction of cocaine and benzoylecgonine from urine using BondElut Certify*, Varian Sample Preparation Products, Harbor City, CA.
132. Bratin, K., Kissinger, P. T., Briner, R. C. and Bruntlett, C. S. (1981) *Anal. Chim. Acta*, **130**, 295.
133. Gupta, R. N. and Steiner, M. (1990) *J. Liq. Chromatogr.*, **13**, 3785.
134. Kramer, P. (1993) *Listening to Prozac: a Psychiatrist Explores Antidepressant Drugs and the Remaking of the Self*, Viking.
135. Caccia, S., Cappim, M., Fracasso, C. and Garatlini, L. (1990) *Psychopharmacology*, **100**, 509.
136. Roettger, J. R. (1990) *J. Anal. Toxicol.*, **14**, 191.
137. Wong, S. H. Y. (1988) *Clin. Chem.*, **34**, 848.
138. Davidson, I. E. and Smyth, W. F. (1977) *Anal. Chem.*, **49**, 1195.
139. Davidson, I. E.(1976) *Proc. Anal. Div. Chem. Soc.*, **13**, 229.
140. Burchfield, H. P. and Storrs, E. E. (1962) *Biochemical Applications of Gas Chromatography*, Academic Press, New York.
141. Wragg, J. S. and Johnson, G .W. (1974). *Pharm. J.*, **213**, 601.
142. Okuda, Y. (1963) *Rev. Polarogr.*, **11**, 197.
143. Oliver, R. and Green, B. N. (1991) *Trends Anal. Chem.*, **10**, 85.
144. Green, B. N., Oliver, R. W. A., Falick, A. M., Shackleton, C. H. L., Roitman, E. and Witkowska, M. E. (1990) in *Biological Mass Spectrometry*, ed. A. Burlingame and J. A. McCluskey, Elsevier Amsterdam, p. 129.
145. Shackleton, C. H. L., Falick, A. M., Green, B. N. and Witkowska, H. E. (1991) *J. Chromatogr.*, **562**, 175.
146. Hemmila, I. (1991) *Applications of Fluorescence in Immunoassays*, Wiley-Interscience, New York.
147. *Application Note 13902603-2*, Wallac Oy, Turku.

Index

absorption maximum 62
accreditation 24
accuracy 24, 25
acetylacetone 46
acid dissolution, of solids 42
aflatoxins 28, 38
air 4, 10, 18, 22, 112, 113
alcoholic beverages 173
aluminium 2, 10, 75
amines, aromatic 20
2-amino-5-chlorobenzophenone 150
p-aminophenol 143
ammonium pyrrolidine
    dithioicarbamate 48, 132
analgesic 143
analyte 10
analytical chemistry 1
analytical method 24, 25
analytical problems 1, 2, 10, 18, 19, 20
analytical scientist 1, 2, 17, 18, 24
analytical system 28, 29
antibiotic 11, 176
antibody 36
anti-cancer drug 20
antimony 9, 10, 130
aqueous environment 10, 98
arsenic 8, 10, 19, 22, 125
arsenic trimethyl 22, 125
ashing 43
aspirin 10, 143
atmosphere 1
atomic absorption spectrometry 10, 22, 28, 42, 48, 89, 102, 112, 127, 130
atomic spectrometry 30
Autoanalyser 71
automation 24, 49, 70, 80, 139
azo dye 123
Azomethine H 58, 108

barbiturate 36
barium 2, 10
benzamide 61
benzo[*a*]pyrene 41
beverage 11
biological fluid 10, 11, 77, 139, 190
biological material 2
blood 4, 20, 44
body fluid 18
boron 2, 8, 10, 58, 108
bromine 2, 10

caffeine 10, 143
calcium 2, 4, 10, 80,
calculation, of analytical result 1, 2, 57
calibration curve 57, 135
capillary zone electrophoresis 10, 24, 58, 65, 109, 152
cephalosporin 11, 152
cerium 2, 5, 10, 83
6-chloro-4-phenyl-2-
    quinazolinecarboxaldehyde 150
choice, of method 1, 18, 19
chromatographic separation 49, 50
clindamycin 176
clinical biochemistry 24, 71
cobalt 2, 6, 10, 94
cocaine 11, 49, 179
codeine 10
coefficient of variation 28
complex matrix 1, 2, 10
complex salt 10, 90
complexometric determination 10, 104
computer 2, 61
confidence limits 37, 52
coning 37
continuous flow visible

spectrophotometry 11, 157
continuous on-line monitoring 75
copper 2, 7, 10, 48, 98
correlation coefficient 58
COSHH Regulations, UK 42, 115
coulometry 18
coupled techniques 50, 102, 112
cross reactivity 36

dacarbazine 20
dairy feedstuff 11, 176
data processing 64
data analysis 64
data recording 64
data storage 64
dc arc emission 127
DDT 31
decentralisation, of measurement 51
definition of problem 19
degradation product 2, 10, 150
Delfia R system 203
derivatisation 20, 113, 165
determinate error 53
2,3-diaminonaphthalene 132
diazonium salt 123
dieldrin 61
differential pulse polarography 21, 127
differential thermal analysis 10, 90
2,3-dihydro-6-
 phenylimidazo(2,1,b)thiazole 20
o,o-dihydoxyazobenzene 71
dimethylglyoxime 46, 96
2,4-dinitrotoluene 184
diode array detector 33, 61
dioxins 29
diphenylamine 184
dissolution test 77
diuretic 47
drinking water 1, 10
drug determination 77

EDTA titrations 10, 45, 80
effluents, industrial 1
electrochemical analysis 50
electrochemical detector 11, 165, 184
electronic materials 63
electrophoretic separation 49
electrospray mass spectrometry 1, 11, 194
environmental samples 11, 170
enzyme linked immunosorbent
 assay(ELISA) 11, 170

erichrome black T 81
erichrome cyanine R 75
error 24, 25
ethyl acetate 2, 11, 173
europium chelates 201
explosives 11

F table 56, 57
F test 54
factory air 1
fish 4, 10, 102
flame emission spectrometry 10, 79
flow injection analysis 74
fluorescein 31
fluorescence detector 11, 71, 166, 187
fluorescence spectrometry 30, 31, 38
fluoride 10, 135
fluoride selective electrode 45
fluorine 2, 9
fluxes, for dissolution of solids 43
fluoroimmunoassay 11, 201
Fourier transform IR spectroscopy 11, 158
foods 18
forensic toxicology 36
formation constant 106
fragmentation pattern 62
fresh water 4
fungicide 21

gas chromatography 10, 11, 18, 20, 41, 60,
 61, 112, 127, 170, 173, 176
gaseous materials 40
gaseous pollutants 40
glucose 16, 73
Good Laboratory Practice 24
gravimetry 1, 10, 18, 51, 83, 96
Griess test 23
gunshot residue 11, 184

haemoglobins 2, 11, 194
hair 10, 22, 125
Health and Safety Executive 116
Henderson–Hasselbach equation 156
hexokinase enzymatic method 73
higher alcohols 2, 11, 173
high molecular weight analytes 2
high performance liquid
 chromatography(HPLC) 1, 10, 11, 20,
 24, 33, 41, 48, 49, 60, 65, 102, 143,
 150, 165, 170, 184, 187, 194

Index

history 1
'Horwitz trumpet' 28
hydride generation 10, 127, 130
hydrochlorothiazide 47
8-hydroxyquinoline(OXINE) 45

immunoassay 36
impurity detection 33
indeterminate error 53
inductively coupled plasma atomic emission spectroscopy(ICPAES) 140
inductively coupled plasma mass spectrometry(ICPMS) 1, 10, 19, 21, 139
industrial products 1
infrared reflectography 17
infrared spectroscopy(IR) 1, 11, 157
inorganic analyte 1, 2, 19
inositol 16
insulin 11, 199
interferences 192
International Standards Organisation (ISO) 24
internal normalisation method 60
internal standard method 59, 173
iodine 2, 10
ion chromatography 127
ion exchange 10, 33, 104
ion selective electrode 10, 135
ion selective electrodes, in automated analysis 71
iridium 22
iron 2, 6, 10, 19, 21, 92
isocyanates 2, 11, 41, 165
isoprene 22
isotopes 4

laboratory information management systems (LIMS) 51
laboratory unit operations 77
lanthanides 83
lead 2, 8, 10
levamisole 20
liquid chromatography 18
liquid materials 39
limestone 4
limit of detection(LOD) 28, 29, 30, 31
limit of quantitation(LOQ) 28, 31
lincomycins A and B 176
linear regression analysis 57
literature search 19

liver 10, 130
lyophilised serum 26

magnesium 2, 4, 10, 71, 80
manganese 2, 6, 10, 90
major constituents 2, 19
masking 45
mass spectrometry(MS) 11, 20, 50, 61, 170, 176
maximum exposure limit(MEL) 116, 170
measurement 1, 10, 18, 50
mercury 7
metabolites 11, 36, 49, 60, 179, 187
methylene blue 157
2-methyl butanol–3-methylbutanol ratio 173
micellar electrophoretic capillary chromatography(MECC) 10, 33, 152
microprocessors 2, 18, 61
migration time 62
minerals 1, 4
mineral water 10, 79
minor constituents 2, 19
molybdenum 2, 5, 10, 89
multielement analysis 2, 139
muscle, mammal 4
myoglobin 194

N-(1-naphthyl)ethylene diamine 123
NAMAS accreditation 24
National Institute of Occupational Safety and Health, USA (NIOSH) 42, 122
natural water 10, 40, 132
neutron activation analysis(NAA) 10, 22, 125
nickel 2, 7, 10, 96
nitrogen dioxide 10, 113
nitroglycerine 23, 184
1-nitroso-2-naphthol 95
nuclear magnetic resonance spectroscopy(NMR) 10, 143

occupational exposure standard(OES) 116
Occupational Safety and Health Administration(OSHA) 42, 122
ores 10, 37, 83, 92
organic analyte 1, 2, 19
organic trace analysis 2
organolead compounds 2, 10, 112
organomercury compounds 2, 10, 102

organometallic analyte   1, 2, 19
oxazepam   2, 10, 150

paracetamol   10, 143
passive smoking   20
peak potential   62
peak purity analysis software   32
peanuts   38
Periodic Table   2, 3
peroxide complexes   84
pesticides   11, 28, 170
pharmaceutical formulation   10, 11, 104, 157
phenacetin   10, 143
o-phenanthroline   46
phosphorus   10
photometry, in automated analysis   71
plants   1, 10, 108
plasma   44
polarography   1, 18
polycyclic aromatic hydrocarbon   41
polymer   11
potassium   4, 10, 79
potentiometry   1, 10, 18, 35
precipitation   45
precision   26, 28
preliminary treatment of sample   18, 42
procaine penicillin G   11, 157
proficiency testing schemes   24, 26
prostate specific antigen   11, 201
prostatic cancer   201
Prozac   11, 60, 187
psicose   16

qualitative analysis   61
quantitative analysis   57
Q tables   52
Q test   51
quartering   37

radiation buffer   80
radiochemical analysis   18, 50
radioimmunoassay   11, 199
recovery   35
redox titration   21, 92
reference materials, certified   24
relative standard deviation   26, 32
reliability, of measurements   51
repetitive analysis, automation of   70
representative sample   1, 18, 37
resolution   32

retention time   62
rice field water   21
river water   10, 40
robotics   51, 76
rock   10, 84
rubidium   2, 10

salicylamide   10
sample sizes   19
sampling   37, 38, 40
sea water   4, 10, 48, 87
sedimentary rock   22
selected ion monitoring(SIM)   176
selectivity   31, 32, 36, 172
    coefficient   33
    constant   35
selenium   2, 9, 10, 35, 36, 132
sensitivity   28
separation   1, 18, 45
serum   11, 44, 47, 73, 187, 201
shoeheel rubber   11, 164
short term exposure limit   116
sickle cell anaemia   194
similarity factor   33
simultaneous equations, solution of   84
single element analysis   2
sodium   2, 4, 10, 79
software, for method development   67
soils   1, 4, 10, 18, 46, 94
solid materials   37, 38
solid phase extraction(SPE)   11, 48, 176, 179
solvent extraction   10, 18, 20, 46, 89, 143
solution to problem   1, 18
sorbent tube   10, 41, 42, 113
speciation   2, 98, 126, 132
spectrofluorimetry   1, 10, 41, 132
spectrometry   50, 84
stability indicating assay   150
stacking, in capillary electrophoresis   109
statistical assessment of measurements   1, 18, 51
standard addition method   58, 134, 193
standard deviation   26, 30
steel   10, 96
stripping voltammetry
    adsorptive   10, 62, 87
    anodic   10, 98
    cathodic   35, 127, 190
sudden infant death syndrome(SIDS)   130
sulphanilic acid   123
system suitability testing   25, 65

Index

t table 53
t test 54
textile dyes 33
thermal methods of analysis 18
thermal desorption gas chromatography 22, 41, 113
thermogravimetry 11, 50, 161
thin layer chromatography 18, 20
thioamides 11, 190
time weighted average (TWA) 116
tin 2, 10
titanium 5, 10, 84
titrimetry 1, 10
toxicity 4
trace constituents 19
trace elements 28, 40
transduction 16
triazines 11, 170
triphenyltin hydroxide 21
tyrosine 199

unit processes 1, 18
urine 11, 36, 49, 71, 179
UV irradiation 88
ultraviolet detector 11, 20, 60, 165
UV visible spectrophotometry 1, 10, 143

Valid Analytical Measurement (VAM) 24
validated analytical methods 24, 25, 65
validation, instrumental 25
vanadium 2, 5, 10, 87
vegetables 4
veterinary pharmaceutical 20
visible spectrophotometry 10, 94, 108, 113, 143
vitamins, water soluble 2, 10, 152
voltammetry 10, 143
volumetric analysis 18, 50

wallpaper 10, 125
water hardness 80
water samples 10, 18, 135
wavelength of absorption 62
workplace atmosphere 11, 40, 42

X-ray fluorescence spectroscopy 10, 125

z score 26
zinc 2, 7, 10, 104